Statistical Analytics for Health Data Science with SAS and R

This book aims to compile typical fundamental-to-advanced statistical methods to be used for health data sciences. Although the book promotes applications to health and health-related data, the models in the book can be used to analyze any kind of data. The data are analyzed with the commonly used statistical software of R/SAS (with online supplementary on SPSS/Stata). The data and computing programs will be available to facilitate readers' learning experience. There has been considerable attention to making statistical methods and analytics available to health data science researchers and students. This book brings it all together to provide a concise point-of-reference for the most commonly used statistical methods from the fundamental level to the advanced level. We envisage this book will contribute to the rapid development in health data science. We provide straightforward explanations of the collected statistical theory and models, compilations of a variety of publicly available data, and illustrations of data analytics using commonly used statistical software of SAS/R. We will have the data and computer programs available for readers to replicate and implement the new methods. The primary readers would be applied data scientists and practitioners in any field of data science, applied statistical analysts and scientists in public health, academic researchers, and graduate students in statistics and biostatistics. The secondary readers would be R&D professionals/practitioners in industry and governmental agencies. This book can be used for both teaching and applied research.

Chapman & Hall/CRC Biostatistics Series

Series Editors

Shein-Chung Chow, Duke University School of Medicine, USA
Byron Jones, Novartis Pharma AG, Switzerland
Jen-pei Liu, National Taiwan University, Taiwan
Karl E. Peace, Georgia Southern University, USA
Bruce W. Turnbull, Cornell University, USA

Recently Published Titles

Quantitative Methodologies and Process for Safety Monitoring and Ongoing Benefit Risk Evaluation
Edited by William Wang, Melvin Munsaka, James Buchanan and Judy Li

Statistical Methods for Mediation, Confounding and Moderation Analysis Using R and SAS
Qingzhao Yu and Bin Li

Hybrid Frequentist/Bayesian Power and Bayesian Power in Planning Clinical Trials
Andrew P. Grieve

Advanced Statistics in Regulatory Critical Clinical Initiatives
Edited By Wei Zhang, Fangrong Yan, Feng Chen, Shein-Chung Chow

Medical Statistics for Cancer Studies
Trevor F. Cox

Real World Evidence in a Patient-Centric Digital Era
Edited by Kelly H. Zou, Lobna A. Salem, Amrit Ray

Data Science, AI, and Machine Learning in Pharma
Harry Yang

Model-Assisted Bayesian Designs for Dose Finding and Optimization
Methods and Applications
Ying Yuan, Ruitao Lin and J. Jack Lee

Digital Therapeutics: Strategic, Scientific, Developmental, and Regulatory Aspects
Oleksandr Sverdlov, Joris van Dam

Quantitative Methods for Precision Medicine
Pharmacogenomics in Action
Rongling Wu

Drug Development for Rare Diseases
Edited by Bo Yang, Yang Song and Yijie Zhou

Case Studies in Bayesian Methods for Biopharmaceutical CMC
Edited by Paul Faya and Tony Pourmohamad

Statistical Analytics for Health Data Science with SAS and R
Jeffrey R. Wilson, Ding-Geng Chen and Karl E. Peace

For more information about this series, please visit: https://www.routledge.com/Chapman-Hall-CRC-Biostatistics-Series/book-series/CHBIOSTATIS

Statistical Analytics for Health Data Science with SAS and R

Jeffrey R. Wilson, Ding-Geng Chen,
and Karl E. Peace

CRC Press
Taylor & Francis Group
Boca Raton London New York

CRC Press is an imprint of the
Taylor & Francis Group, an **informa** business

A CHAPMAN & HALL BOOK

Designed cover image: https://www.shutterstock.com/image-photo/doctor-hand-working-laptop
-computerconcept-online-1868420722, edited by Anne Rubio

First edition published 2023
by CRC Press
6000 Broken Sound Parkway NW, Suite 300, Boca Raton, FL 33487-2742

and by CRC Press
4 Park Square, Milton Park, Abingdon, Oxon, OX14 4RN

CRC Press is an imprint of Taylor & Francis Group, LLC

ISBN: 978-1-032-32562-0 (hbk)
ISBN: 978-1-032-32569-9 (pbk)
ISBN: 978-1-003-31567-4 (ebk)

DOI: 10.1201/9781003315674

Typeset in Palatino
by Deanta Global Publishing Services, Chennai, India
Access the Support Material: www.public.asu.edu/~jeffreyw

Contents

Preface

In the statistical analysis of data, it is often of key interest to evaluate the impact of a covariate or covariates on the response variable through a statistical model. The type of statistical model used in this process depends a great deal on the way the data were obtained and the type of variable (measure) being used as the response. In this book, we focus on models with one categorical response or one continuous response. In the first 8 chapters, we focus on data obtained based on independent observations and one or more covariates. In Chapters 9, 10, and 11, we introduce models applicable when the observations are correlated.

The focus of this book is the modeling of response outcomes with categorical and or continuous factors. We strive for simplicity and ease in the explanation of statistical terms and terminologies. We refer to practical examples whenever there is an opportunity to simplify a concept or to bring theory into practice.

We use several real-world datasets to demonstrate the models fitted throughout the book. They are data encountered over the last few years of our research and teaching of statistics at the undergraduate level at our universities. The aim of this book is to concentrate on using marginal models. These are models that address the overall mean of the response variable.

The examples in this book are analyzed whenever possible using SAS and R code. The code is provided. The SAS and the R code present outputs which are contained in the text with partial information at times. The completed datasets can be found at www.public.asu.edu/~jeffreyw. We provide several examples to allow the reader to mimic some of the models used. The chapters in this book are designed to help guide researchers, practitioners, and nonstatistical majors to model outcomes using covariates.

The book is timely and has the potential to impact model fitting when faced with data analyses and a list of practical questions. In an academic setting, the book could serve as a teaching textbook to understand the opportunities the analysis of independent observations allows. It contains, though to a lesser extent, an introduction to modeling correlated observations for beginners and those practitioners. The book is useful to those who desire some intermediate outlook on the discipline or for those seeking degrees in related quantitative fields of study. In addition, this book could serve as a reference for researchers and data analysts in education, the social sciences, public health, and biomedical research.

Readings

The book is comprised of different opportunities for readers. Those interested in just one response and one covariate can read Chapters 1, 2, 3, and 4. Those interested in statistical models with a multiple variable setting can read Chapters 5, 6, 7, and 8. Those who want to be introduced to modeling correlated observations (binary response or continuous response) can read Chapters 9, 10, and 11.

Acknowledgments

The authors of this book owe a great deal of gratitude to many who helped in the completion of the book. We have been fortunate to work with a number of graduate students at Arizona State University. Many thanks to the staff in the Department of Economics and the computing support group in the W. P. Carey School of Business, College of Health Solutions. To everyone involved in the making of this book, we say thank you!

We dedicate this to our students, present and past. Their insight had a great deal to do with the material covered in this book.

Jeffrey R. Wilson, Ding-Geng Chen, and Karl E. Peace

Author Biographies

Jeffrey R. Wilson, Ph.D., is Professor in Biostatistics and Associate Dean of Research in the Department of Economics at W. P. Carey School of Business, Arizona State University, USA.

Ding-Geng Chen, Ph.D., is Professor and Executive Director in the Biostatistics College of Health Solutions at Arizona State University, USA.

Dr. Karl E. Peace is the Georgia Cancer Coalition Distinguished Cancer Scholar (GCCDCS), Senior Research Scientist and Professor of Biostatistics in the Jiann-Ping Hsu College of Public Health (JPHCOPH) at Georgia Southern University (GSU), USA. He was responsible for establishing the Jiann-Ping Hsu College of Public Health – the first college of public health in the University System of GA (USG). He is the architect of the MPH in Biostatistics – the first degree program in Biostatistics in the USG and Founding Director of the Karl E. Peace Center for Biostatistics in the JPHCOPH. Dr. Peace holds a Ph.D. in Biostatistics from the Medical College of Virginia, an M.S. in Mathematics from Clemson University, a B.S. in Chemistry from Georgia Southern College, and a Health Science Certificate from Vanderbilt University.

List of Abbreviations

Abbreviations
CI   Confidence interval
GLM   Generalized linear model
GLMM   Generalized linear mixed model
MLE   Maximum likelihood estimate
SE   Standard error

Definitions
Response   denotes a measurement on different subjects
Observation denotes the measurement on a particular subject/unit at different times

1

Survey Sampling and Data Collection

1.1 Research Interest/Question

Sampling is a scientific approach that enables us to describe the characteristics of a population without having to audit everyone in the population (census). One often asks:

1. What size should the sample size be?
2. Do I have a sufficient number of sampling units?
3. For a given sample size, how accurate will my answers be?

While the desire is usually to identify characteristics of the entire population, it is usually impossible to assess the entire population because of time and costs – including personnel. Thus, one must rely on a sample from the population (i.e., subset). The statistical models applied to the sample data are important, as these results are used to make inferences to the sampled population. In this chapter, methods for obtaining a useful and appropriate subset of the population are presented. This chapter discusses sample surveys and introduces different sampling techniques. This chapter also discusses the major issues that a researcher faces when selecting an appropriate sample. The steps in a survey can be summarized as follows:

1. Specify a sampled population based on one's judgment;
2. Use a probability sampling scheme to obtain a sample;
3. Fit models to the sample data;
4. Make inferences to the sampled population.

Why sampling? Sampling is a process of selecting a subset of the population. The use of the sample helps to minimize the costs of collecting, editing, and analyzing data for the entire population. Thus, a sample allows the researcher to economize the data generation process with a trade-off of having less information.

DOI: 10.1201/9781003315674-1

1.2 Some Basic Terminology

Population is the set of units of interest. The population must be fully defined. We make reference to two types of populations: the target population and the sampled population (Figure 1.1). The **sampled population** is the population from which one takes the sample. It is the group to which we make inferences. While the **target population** is the population of interest, we usually work with the sampled population, hoping that it mirrors the target population as closely as possible. In some cases, the target and the sampled population may be identical. In other cases, they may have no elements in common.

Figure 1.1 shows the steps in the process. Through judgment, the sampled population is determined to be the population of interest. Several reasons may hinder one from reaching the target population. From the sampled population a subset is taken. One hopes that the subset is taken in such a way that it is a good representation of the characteristics of the population. The sample is used to fit statistical models, to estimate parameters, and to make inferences back to the sampled population.

> **The target population** is the population of interest. It must be specified, otherwise, it is impossible to know to which population of units inferences are made. In a more complicated structure, we may have a population of primary units and a population of secondary units within the primary units.
>
> **Census** is the act of sampling the entire population. The process of a census takes a considerable amount of time and other resources.

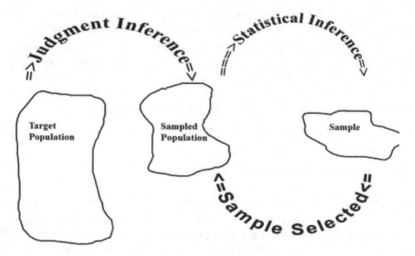

FIGURE 1.1
Graphical display of steps in survey

Major advantages of sampling rather than complete enumeration are savings in cost and in time.

Sample is a subset of the sampled population. One observes or obtains a sample with the main purpose of obtaining information about the population of interest. In some health studies, for example, a sample of people may be observed for exposure to various risk factors. The ability to infer from the sample to the population depends largely on the sample. Therefore, the accuracy of the conclusions will depend on how well the sample was collected. Also, it depends on how representative the sample is of the population.

Sampling frame is a list of sampling units. It provides a list from which the experimenter usually plans to draw a sample. It is not uncommon to find it impossible to identify an accurate or up-to-date sampling frame. A survey based on a random sampling procedure will almost always provide a sample that is representative of the population under study. As such, it allows the findings to be generalizable to the entire population. The term "representative", as it is commonly used, means that the sample resembles the population in some key ways (Babbie 1990).

Parameter is a value obtained from the population. It is usually unknown.

Statistic is a value obtained from the sample. It provides an estimate of the parameter.

In studies of cause and effect, a **confounding variable** is any other variable that can influence both cause and effect. Confounding variables can play a major role in analyzing cohort studies. They are often seen as an extra predictor that has a hidden impact on effect (and cause). An example that is often given is high incidences of sales of ice cream cones and sunburn simultaneously in New York City. Does eating an ice cream cone cause sunburn? No, but both variables are influenced by high temperatures. So, the temperature would be a confounding variable in this example.

1.3 Sample Properties

A **representative sample** of a population is dependent upon the accuracy of the sampling frame used.

Question: What determines an adequate sample?

Answer: The primary concern in selecting an appropriate sample is that the sample should be representative of the population. The desire is that all variables of interest should reflect the same distribution in the sample as well as in the population from which the sample is chosen. However, one usually does not know enough information about the population to know if the sample and the population are mirror images. However, one must do all things possible to accomplish the best opportunity for representativeness. Before a sample is drawn, one should fully describe the population. In a population survey, this requires having a list (sampling frame) of all the individuals in the population. A probability sampling method is developed to draw a sample in such a way that one can assure the representativeness of the key characteristics of interest. The completeness and the accuracy of the sampling frame are essential components for the study to be successful. One of the major flaws in many survey research studies is a biased selection of the sampling units.

Example from *Tricks and Traps* (Almer 2000): In 1936, Literary Digest sent out 10 M questionnaires regarding the election of Roosevelt versus Landon. The response rate was 24% or 2.4 M, with a clear majority preferring the Republican, Alfred Landon over Democrat Theodore Roosevelt. The poll was far from a correct prediction, as Roosevelt won by a landslide.

  Question: Why was there such a big difference between the poll results and the actual results?

  Answer:
  - The sampling frame was the mailing list of Literary Digest readers;
  - Wealthy individuals tended to be Republican and members of the Digest;
  - The poll was biased;
  - The high nonresponse rate.

  Size of the sample: Once a sampling frame has been identified, one needs to have a method for selecting the units from this frame. Two issues are important:
  1. How should the individual units be selected? (sampling scheme)
  2. How large of a sample should be selected? (sample size)

One of the most difficult decisions facing the researcher is how large the sample should be. Two common approaches are employed in research studies to address the empirical and the analytical.

  1. The **empirical** approach involves using sample sizes that have been used in similar studies. There is no scientific basis for this approach. It depends on the success of previous studies. The generalization

depends on how similar the present study is in its scope (objectives, design, study population, etc.). Thus, there are limits on the generalization;

2. The **analytical** (scientific) approach is to use a proportion of the population and the desired margin of error while considering the cost (Babbie 1990).

Bias in sampling: There are key sources of bias that may affect the selection of a sample. The following conditions will introduce bias:

1. *Bad protocol*: Lack of a defined protocol or unable to follow;
2. *Unable to locate*: Sampling units are difficult to reach and omission;
3. *Replacement*: Replacing selected units for other reasons;
4. *Nonresponse*: High nonresponse rates;
5. *Improper sampling frame*: An out-of-date sample frame (Schaeffer et al. 1995).

Reliability refers to the consistency or stability of measurement. Can the measure or other forms of observation be confirmed by further measurements? If you measure the same thing, would you get the same score? An instrument is said to be reliable if it accurately reflects the true score, and thus minimizes the error component. Reliability and variance behave as opposites; e.g., low reliability implies high variance.

Validity refers to the suitability or meaningfulness of a measurement. Does this instrument accurately describe the construct you are attempting to measure? In statistical terms, validity is analogous to unbiasedness (*valid = unbiased*). Validity indicates how well an instrument measures the item that it was intended to measure.

Accuracy is how close a measured value is to the actual (true) value. It refers to how far away a particular value of the estimate is, on average, from the true value of the parameter being measured.

Precision is how close the measured values are to each other.

1.4 Probability Sampling Design

It would generally be impractical and too costly to study an entire population. However, probability sampling is a method that allows researchers to make inferences about a population parameter without having to take the entire population (census) as discussed in Levy and Lemeshow (1999). Reducing the number of units in a study minimizes the cost and the workload and may make it easier to obtain high-quality information. However, there is a trade-off. The science of sampling is to balance between having a large enough sample size with enough power to detect true differences and the cost of sampling. It is important that the subjects chosen are representative of the entire population. This is often complicated when there are hard-to-reach groups.

Probability versus nonprobability samples: There are two types of samples, probability and nonprobability samples. The difference between probability and nonprobability sampling is that nonprobability sampling does not involve random selection, but probability sampling does. Does that mean that nonprobability samples are not representative of the population? Not necessarily! However, it does mean that nonprobability samples cannot depend upon the rationale of probability theory. At least with a probability sample, one knows the probability with which the sample represents the population. Also, one can obtain confidence intervals (a range of possible values) for the parameter.

However, with nonprobability samples, one may or may not represent the population adequately, and it is difficult to know to what degree one has done so. In general, researchers prefer probability random sampling methods to nonprobability methods. The probability sample methods are more accurate and reliable. However, in applied research, there may be circumstances when a probability sample is not feasible, not practical, or theoretically convenient.

There are four basic random sample designs (schemes). In practice, there may exist a combination of these basic designs; in such a case, we refer to these as complex schemes. The size of the sample will play a role in the choice of the sampling designs.

1.4.1 Simple Random Sampling Design

In this case, each individual is chosen entirely by chance, and each member of the population has an equal chance, or probability, of being selected. One way of obtaining a random sample is to give each unit or subject in a population an identification number, and then include numbers in the sample according to those from a table of random numbers. The individual is the sampling unit. A simple random sample is the most common and the simplest of the probability sampling methods. The simple random sample has several advantages. It is simple to administer and has a good chance of being representative of the population. The analysis of the resulting data provides a variance that is straightforward to compute. The disadvantage is that the selected sample may not be truly representative of the population, especially if the sample size is small and the population is heterogeneous. Also, it requires a sampling frame or a list of the sampling units.

Question: Is a simple random sample always appropriate for all cases?

Answer: Probably not!

Question: Do you have a scenario that supports your response?

Answer: Imagine one wishes to estimate the mean blood pressure for your class. The class has 20 females and 80 males. It is possible to obtain a random sample of 20 consisting of 10 females and 10 males. Men usually have higher blood pressure than females. The population consists of a larger proportion [80/(80+20) = 80%] of males, but the sample has an equal proportion of males and females. This disproportion gives reason for us to be concerned.

Question: How can one avoid such or similar scenarios?

Answer: Use a stratified random sample.

1.4.2 Stratified Random Sampling Design

A stratified random sample consists of segmenting the population into subgroups (strata). From each stratum, one takes a simple random sample of units. A stratum consists of units with similar characteristics. The segmentation depends on what you are studying. Within each stratum, units are more alike than between strata. The individual is the sampling unit. A stratified random sample allows one to have a certain representation in each stratum (subgroup). It ensures that a particular group or category of units is represented in the sample. If one wanted to ensure that the sample is representative of the population frame, then one can employ a stratified sampling scheme. Stratified sampling is advantageous when the information about the distribution of a particular characteristic is known. One selects simple random samples from within each of the subgroups defined. The fact that the population is stratified is taken into account at the computation stage. It comes into the analysis through the variance of the estimators.

Example: Let us return to the blood pressure example: In this example, one can take a stratified random sample with gender providing the strata – males and females. If one took 20/(20+80) = 20% of the sample size from the female stratum and 80/(20+80) = 80% from the male stratum, one assures that the sample is representative of the population with respect to gender.

Advantages:
- Stratified sampling reduces potential confounding by selecting homogeneous subgroups;
- Stratified random sampling is more accurate than simple random sampling with similar sample sizes;
- Stratified random sampling provides more information about subgroups.

Disadvantages:
- Stratified random sampling is more costly than simple random sampling.
- **Question:** The stratified sampling scheme is more expensive as one has to travel to diverse areas to sample units from each

stratum. In these surveys, the cost can be exorbitant, as it often requires one to carry out interviews with individuals scattered across the country. How can one keep the cost down and yet avoid some of the problems encounter with simple random sampling?

- **Answer:** Use a cluster random sample.

1.4.3 Cluster Random Sampling Design

A cluster random sample consists of segmenting the population into clusters (groups) conveniently. Then, one takes a simple random sample of the clusters (groups). It is a method frequently employed in national surveys. Cluster sampling results in individuals selected in clusters or families or census tracts or geographical batches.

Cluster sampling in many surveys is done on large populations, which may be geographically quite dispersed. In such cases, clusters may be identified (e.g., households) and random samples of clusters are included in the study; then every member of the cluster is a part of the study. This introduces two types of variations in the data – between clusters and within clusters – which are taken into account when analyzing data. In a clustered sample, clusters (subgroups of the population) are used as the sampling unit, rather than individuals. All members of the cluster are included in the study. The sampling scheme dictates the variance used in constructing the test statistic or confidence interval. The two components of the variance, within clusters and between clusters, result in less accurate estimates than with simple random sampling design of similar sample sizes.

Advantage:

- Cluster random sampling is less costly than simple random sampling. The interviewer goes into a cluster and does not have to be dispersed over several areas.

Disadvantage:

- Cluster sampling is less accurate as is the case in simple random sampling. It consists of correlated observations on account of the grouping brought in by the clustering.

Question: The clustered sampling scheme is less accurate but less costly. Is there a sampling scheme with less cost but is easy to administer?

Answer: Yes, the systematic random sample.

1.4.4 Systematic Random Sampling Design

Systematic random sampling consists of having the units lined in some uniform system, and every s^{th} unit is taken repeatedly. It is unique in that the first individual, *s*, is picked using a random number to dictate. The subsequent units are selected (i.e., every sth unit) based on the predetermined sample size. In reality, there is only one random event – the first. In fact, this probability sampling design of a sample of *n* selected units consists of one random event as opposed to a simple random sample where there are *n* random events. Thus, systematic sampling is often convenient and easy to administer. However, depending on the ordering of the sampling frame, it may also lead to bias. There is one sampling unit. The sample is systematic with one random event. It creates a problem when the variance is required to construct a test statistic or confidence interval, as one cannot estimate the variance of one random sampled unit. One cannot estimate the variance of one observation. However, it is common practice to obtain a systematic sample and then treat it as a simple random sample in order to obtain an estimate of the variance (Kalton 1983).

1.4.5 Complex Sampling Design or Multi-Stage Sampling

A cluster random sampling scheme has clusters as sampling units. Sometimes these clusters (*primary units*) are too large to sample the entire subgroup. One takes a sample of the units from the selected cluster (primary units) at random. This provides primary and secondary units. Many studies, especially large nationwide surveys, incorporate different sampling methods for different groups at several stages (Levy and Lemeshow 1999).

Advantages:
> The main advantages of a complex sample in comparison with a simple random sample are:

- Complex sampling does not require a complete sampling frame;
- Complex sampling is more economical and practical;
- Complex sampling guarantees a representative sample of the population, once they are formed strategically;
- Complex sampling articulates the process systematically.

Disadvantage:

- Complex sampling is generally less efficient than simple random sampling. It provides estimates with less precision than with simple random sampling.

1.5 Nonprobability Sampling Schemes

Nonrandom sampling scheme is not commonly used in quantitative social research. However, it is used increasingly in market research surveys. The technique most commonly used is called quota sampling.

Quota sampling scheme is a technique for sampling whereby the researcher decides in advance on certain key factors from the strata. Interviewers are given a quota of subjects of a specified type to interview. The difference between a stratified random sample and a quota sample is that the units/respondents in a quota sample are not randomly selected within the strata. The respondents are subjectively selected for convenience or familiarity. It is conducted without the use of random sampling. Therefore, it is not valid to extrapolate or generalize the findings to a larger population. There are several flaws with this method, but most importantly, it is not truly random. So, one cannot find confidence intervals or construct test statistics, as there are no true variance estimates.

Convenience sampling or **opportunistic sampling scheme** is a relatively easy method of sampling. The participants are selected subjectively in the most convenient way. The participants are allowed to choose and volunteer to take part. The interviewees are easily accessible, thus the name "convenience sampling". The technique is generally not respected by survey sample researchers; however, it is an acceptable approach when using a qualitative design since extrapolation is not a primary concern of qualitative researchers. These researchers will defend that they get good results, while others will declare that the data are severely biased. The biasedness is due to self-selection, as those who volunteer to take part in the survey may be different from those who choose not to take part.

Snowball sampling scheme is where, once a subject is chosen, the sampled subjects are asked to nominate further subjects who can be interviewed. Thus, the sample increases in size, much as a rolling snowball, as the saying goes.

1.6 Surveys

Surveys in public health are used to determine different health choices or certain behavioral habits of individuals. For example, one may wish to know how a certain population behaves in terms of smoking, exercise, seatbelt or helmet use, or physical exams, etc. A survey is a useful tool designed to investigate these and other topics. Surveys often use questionnaires as tools for data collection. The biggest challenge in survey research is the response rate. Nonresponse is known to create certain challenges, including bias (Aday and Cornelius 2006).

Cross-sectional surveys give a view of the population at a particular point in time. They provide a snapshot of what is happening in the population. They provide a descriptive look at a subset of the population.

Longitudinal surveys provide information over time. These surveys sometimes last for months or years. Longitudinal surveys usually take one of two forms:

1. **Cohort surveys** – follow the same group of individuals over time;

2. **Trend surveys** – take repeated samples of different people each time but always use the same core questions.

1.6.1 Studies

Observational studies involve cases where the researcher simply observes the process. The observer has no input into the design. This differs from an experimental study, where the researcher controls and oversees the data generation process. When used appropriately, observational studies allow investigation of incidence (i.e., the number of new cases in a fixed time divided by the number of people at risk). If the period of study is one year, one can report the annual incidence.

In addition, observational studies are often the only option in research methods. In particular, observational studies are preferred when a randomized controlled trial would be impractical or unethical (Farmer et al. 1996). An observational study may be a cohort study, a cross-sectional study, or a case-control study.

1.6.2 Cohort Study

In **cohort studies**, subjects are followed over time. Subjects are selected before the outcome of interest is observed. These may be **prospective** or **retrospective**. They may be cheaper and quicker. At times, they are used to study incidence, causes, and prognoses. They are preferred for determining the incidence and the natural history of a condition. Researchers often use such studies to distinguish between cause and effect. **Prospective cohort studies** consist of a group of subjects chosen who do not have the outcome of interest (for example, obesity). The investigator measures a cadre of covariates on each subject that might be relevant to the development of the condition. Over a period of time, subjects in the study are observed to see whether they develop the outcome of interest (i.e., obesity). Prospective studies are expensive and often take a long time for sufficient outcome events to occur to produce meaningful results. **Retrospective cohort studies** usually consist of study data already collected for other purposes. The methodology is the same, but the study is performed post hoc. The cohort is "followed up", retrospectively. The study period may be many years, but the time to complete the study is only as long as it takes to collect and analyze the data.

1.6.3 Cross-Sectional Studies

Cross-sectional studies are used to determine prevalence. They are relatively quick and easy. However, they do not allow the distinction between cause and effect. The studies are useful for identifying associations. The findings are usually void of explanation. However, it is a challenge to study rare conditions efficiently using cross-sectional studies. The fact is that there may not be enough representation in the sample.

1.6.4 Case-Control Studies

Case-control studies compare groups retrospectively. While cross-sectional studies consider each subject at one point in time, case-control studies look back at what has happened to each subject. Subjects are selected specifically based on the outcome of interest. Subjects with the outcome of interest are matched with a control group (a group of subjects without the condition of interest, or unexposed to or not treated with the agent of interest). Case-control studies are also called retrospective studies.

1.7 Key Terms in Surveys

What determines an adequate sample? The primary concern in selecting an appropriate sample is that the sample should be representative of the population. The desire is that all variables of interest should reflect the same distribution in the sample as well as in the population from which the sample is chosen. However, one usually does not know enough information about the population to know if the sample and population are mirror images. However, one must do all things possible to accomplish the unknown state of representativeness. Before a sample is drawn, one should describe the population to the best information possible. In a population survey, this requires having a list (sampling frame) of all the individuals in the population. A probability method is developed to draw a sample in such a way that one can assure the representativeness of the various characteristics of interest. The completeness and accuracy of the sampling frame are essential for the study to be successful. One of the major flaws in many research studies is a biased selection of the sampling units.

    Statistical hypothesis: A statement about the parameters describing a population. It is a statement about something you do not know. It is not about a statistic. We know statistical values, but we do not know parameter values. There is the null hypothesis and the research or alternative hypothesis. The null hypothesis assumes the event will not occur or there is no association between the measures under consideration. It has no relation with the outcomes under consideration. It is the "I have to do nothing" hypothesis. It is symbolized by H_0. The research or alternative hypothesis is the logical opposite of the null hypothesis. It suggests that there is a relation, and it requires that some action be taken when this hypothesis is supported. It is symbolized by H_1.

Type I error: In the execution of a survey for estimation or statistical inferences, it is known that despite all the care and attention that one exercises, there is still the possibility that after statistically significant results are declared (rejection of the null hypothesis of no difference), one learns later that the results are not statistically significant. So an error in the decision was made. Such an error is called a Type I error – or a false positive.

Significance level: The probability of the type I error is called the significance level. The significance level is symbolized as alpha (α). It represents the client's comfort level to the researcher.

Type II error: If one declares the results as not significant, and later it turns out they are significant, an error in a decision was made. One refers to the error as the type II error or false negative.

Beta, β: The probability of type II error is symbolized as β.

Type I versus Type II error: Both type I and type II errors are decision errors that can occur when analyzing data collected from having conducted an experiment for the purpose of making a decision about some hypothesized event (called the "null hypothesis" and denoted by H_0). The experimenter chooses the probability of the type I error before experimentation begins. Most often it is chosen as 0.05. If one is not concerned about limiting the probability of the type II error, one does not have to choose its magnitude (prior to beginning experimentation). If one is concerned about limiting the magnitude of the type II error, then β is specified, realizing that the smaller it is the larger will be the sample size for the experiment.

The **power** of a test $(1-\beta)$ is the complement of the magnitude of the probability of type II error. It is the probability that the decision maker rejects H_0 when it is not true; i.e., probability of rejecting H_0 when it should be rejected. It is the probability that we concluded significance and later found that it was the correct decision.

Sampling error: When sampling from a population, one is interested in using data contained in the sample to provide an estimate of a characteristic of the population, e.g., it is logical to use the mean of a random sample (denoted by $\bar{x}$) obtained from the sampled population as an estimate of the population mean (denoted as μ). If the sample mean is not identical to the population mean, sampling error has occurred.

Margin of error is a common summary of sampling error that quantifies uncertainty about the data used from the survey result. An important factor in determining the margin of error is the size of the sample. Larger samples are more likely to yield similar results (statistics) close to the target population (parameter) quantity and thus have smaller margins of error than more modest-sized samples.

> **Example:** Consider a simple random sample of n independent observations obtained from a finite population of N dichotomous outcomes (by dichotomous, we mean that there are two possible outcomes, e.g., say success, failure), then the
>
> $$Margin\, of\, Error = \sqrt{(N-n)/(N-1)} \times 1.96 \times p_{success} \times (1 - p_{success})/n,$$
>
> where N is the population size and n is the sample size, and $P_{success}$ is the number of successes in the sample divided by the size of the sample.

Three things affect the margin of error: sample size, the type of sampling design, and the size of the population.

> **Sample sizes and margin of error**: It is common to have as an objective the purpose of obtaining an estimate of a population parameter with a certain margin of error. In such a case, the size of the sample depends on several factors:
>
> 1. What is the scale of the measure of interest? binary versus continuous;
> 2. What is the underlying probability distribution of the variable of interest? If it is a binary variable, then it is usually assumed to be binomial when one wants to estimate the proportion. If it is continuous, it is usually assumed that it follows the normal distribution, and therefore, one wants to estimate the mean;
> 3. What is the sampling distribution of the statistic? This applies to the theoretical behavior of the statistic, if repeatedly sampled. This distribution is called the sampling distribution. In calculating sample sizes, it is often assumed that the sampling involves simple random sampling. However, often is the case that the sampling design is much more complicated, so that the variance will be more involved;
> 4. How accurate do you want the results to be? One is interested in obtaining an estimate as close to the population parameter as possible. Therefore, one needs a measure of the difference

between the estimate and the unknown population value. We use the standard error to assist us. The standard error comes from the sampling distribution of the statistic being used. If the sampling is done properly *(with appropriate design)*, one can predict the distribution of statistics, and based on this, one can estimate how close to the population value the sample value lies.

Sample sizes and power: It is a common purpose of analytical studies to test hypotheses. As such, one needs to determine the sample sizes required and the magnitudes of the type I and type II errors.

- The larger the sample size, the more information one obtains – but the higher the cost;
- Limiting the probabilities of the two types of errors is not a "one size fits all". One has to determine the measure involved (a proportion, a sample mean, an estimate of relative risk, odds ratio, etc.);
- The method whereby the data are collected, the test statistic used to make a decision, and the sampling distribution of the test statistic under the null hypothesis;
- If one knows any three of these four components, i.e., sample size, significance level, power, and sampling design (which provides the variance, σ^2), then one can compute the fourth;
- A summary is given in Figure 1.2.

The sample size in a hypothesis test depends on power, type I error risk level, and variance, once the scale of the response variable is identified and the related test statistic is obtained. Figure 1.2 shows the relation. Power is a

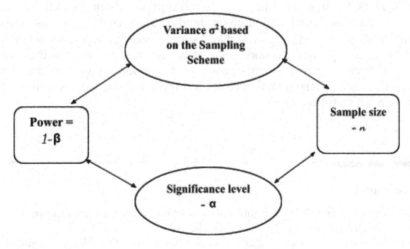

FIGURE 1.2
Relation of power, TYPE I risk level, variance, and sample size

terminology required in hypothesis testing. It tells the probability of declaring significance when, in fact, there is significance. Readers are encouraged to refer to Cohen (1988), for comprehensive discussions on sample size and statistical power, and to Chen et al. (2017) on how to calculate the sample size using statistical software R and SAS.

1.8 Examples of Public Health Surveys

1.8.1 Los Angeles County Health Survey (2015)

(http://publichealth.lacounty.gov/ha/LACHSBackMeth2015.htm)

The Los Angeles County Health Survey (LACHS) provides information about access to health care, health care utilization, health behaviors, health status, and knowledge and perceptions of health-related issues in LA County. The large sample size allows the survey to provide estimates of the health of the county population and of people residing in the county's different geographic regions. Data are collected from households of all educational and income levels. This includes residents living below the federal poverty level.

1.8.2 National Crime Victimization Survey

(https://www.bjs.gov/index.cfm?ty=dcdetail&iid=245)

The Bureau of Justice Statistics (BJS) National Crime Victimization Survey (NCVS) is the key source of information on the nation's crime victimization. Annually, data are collected from a representative sample of about 90,000 households across the nation. It comprises about 160,000 persons. The NCVS collects information on nonfatal personal crimes, such as rape or sexual assault, robbery, aggravated and simple assault, and personal larceny and household property crimes, such as burglary, motor vehicle theft, and other theft. Survey respondents provide information on age, sex, race and Hispanic origin, marital status, education level, income, and whether they experienced victimization.

References

Aday, L.A., Cornelius, L.J.: *Designing and Conducting Health Surveys: A Comprehensive Guide* (3rd ed.). Jossey-Bass, San Francisco (2006)

Almer, E.C.: *Statistical Tricks and Traps: An Illustrated Guide to the Misuse of Statistics.* Pyrczak Publishing (2000)

Babbie, E.R.: *Survey Research Methods*. Belmont, Wadsworth (1990)

Chen, D.G., Peace, K.E., Zhang, P.: *Clinical Trial Data Analysis Using R and SAS* (2nd ed.). Chapman and Hall/CRC, Roca Raton (2017)

Cohen, J.: *Statistical Power Analysis for the Behavioral Sciences* (2nd ed.). Lawrence Erlbaum, Hillsdale (1988)

Farmer, R., Miller, D., Lawrenson, R.: *Epidemiology and Public Health Medicine* (4th ed.). Blackwell Science, Oxford (1996)

Kalton, G.: *Introduction to Survey Sampling*. Sage, Newbury (1983)

Levy, P.S., Lemeshow, S.: *Sampling of Populations. Methods and Applications* (3rd ed.). Wiley, New York (1999)

Schaeffer, R.L., Mendenhall, W., Ott, L.: *Elementary Survey Sampling*. Duxbury Press, Washington, DC (1995)

2

Measures of Tendency, Spread, Relative Standing, Association, and Belief

2.1 Research Interest/Question

When one obtains data, the intention is to reveal the characteristics of the distribution of certain variables or associations among certain variables to be used in policy making or in model building. As such, it is important to have certain information to explain the pattern of these variables. The variables in a dataset, usually contain one or more variables of interest (response variable). The distributions of these variables have certain properties; there is a mean, a variance, skewness, and kurtosis to name a few. One usually starts with the mean but can start with other measures of central tendency to address certain questions about the variable of interest. More so, the parameters of that distribution, in particular the mean, may be impacted by a set of covariates. The relationship of these covariates to the mean parameters of the distribution of the response variable is of great interest. In this chapter, we make use of the following variables: body size, glycohemoglobin, and demographic information from the National Health and Nutrition Examination Survey (NHANES) 2009–2010 which are described.

(https://www.cdc.gov/visionhealth/vehss/data/national-surveys/national-health-and-nutrition-examination-survey.html).

Some basic measures of certain variables and their association with other variables are provided. As an example, one may be interested in Body Mass Index (BMI) and ask the following questions:

1. What proportion of the population is obese?
2. On average, is BMI distributed the same for males as it is for females?

2.2 Body Size, Glycohemoglobin, and Demographics from NHANES 2009–2010 Data

Statistical measures are obtained using variables in the NHANES data. The NHANES data are unique, as these data combine personal interviews with standardized physical examinations and laboratory tests to gather information about illness and disability. The interviews are conducted in the home and at a mobile examination center. It is reported that approximately 5,000 people per year participate in NHANES as a representation of the entire US population. A subset of these data is presented in Table 2.1. Table 2.1 contains ID (i.e., SEQN-respondent sequence number), gender, TX (on insulin or diabetes meds), BMI, leg, waist, and obesity (i.e., BMI above 29.9), among other information for each survey participant.

An examination of these data suggests that the mean of the BMI distribution for males differs from that of females (see Figure 2.1). However, this

TABLE 2.1

Subset of NHANES Dataset

SEQN	Gender	TX	BMI	Leg	Waist	Obese
51624	Male	0	32.22	41.5	100.4	1
51626	Male	0	22	42	74.7	0
51628	Female	1	42.39	35.3	118.2	1
51629	Male	0	32.61	41.7	103.7	1
51630	Female	0	30.57	37.5	107.8	1
51633	Male	0	26.04	42.8	91.1	0
51635	Male	1	27.62	43	113.7	0
51640	Male	0	25.97	39.8	86	0
51641	Male	0	16.6	39.2	63.6	0

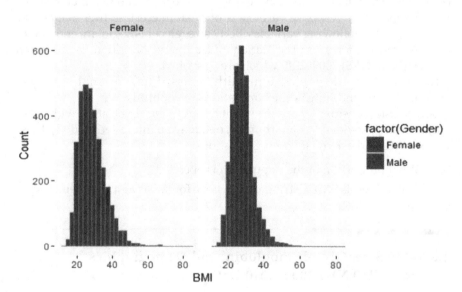

FIGURE 2.1
Graphical distribution for BMI by gender

observation is based on a graphical representation. It does not provide any statistical inferential evidence. One needs to use a statistical test to determine whether the difference between males and females in terms of BMI is large enough to be statistically significant.

2.3 Statistical Measures of Central Tendency and Measures of Spread

2.3.1 Variables

In most studies involving statistical analyses of data, it is important to distinguish between the response variable (i.e., variable of interest) and other (input) variables (covariate, driver, factor, or explanatory variables) that may be related to the response variable. Once this distinction is made, one needs to know the scale of each variable employed. For the response variable, one needs to determine which distribution best describes the responses. For the input variable, one needs to know the scale on which this variable is measured. Variables are classified into one of four types as they pertain to the scale employed.

A **qualitative or categorical** variable is either nominal or ordinal. A **nominal-scale** variable is a variable whose values are categories without any numerical ranking, such as race, state, or region. Nominal variables with only two mutually exclusive categories are called binary or dichotomous or indicators. Examples are alive or dead, cancer or no cancer, and vaccinated or unvaccinated. An **ordinal-scale** variable has values that can be ranked but are not necessarily evenly spaced, such as grades A, B, C, D, and F, or progression of cancer (stages I, II, III, and IV).

A **quantitative or continuous** variable is either interval or ratio. An INTERVAL-SCALE variable is measured on a scale of equally spaced units but without a true zero point (think of a base value), such as date of birth. A RATIO-SCALE variable is an interval variable with a true zero point, such as height in centimeters or duration of illness.

2.3.2 Measures of Central Tendency

The distribution of a variable consists of several parameters (a parameter is a population value). One basic but important parameter is the central location of the values. A measure of central tendency provides a single value that gives an idea of the centering of the distribution of values. It is an essential representation of the distribution of observations. Such a measure includes the mode, median, arithmetic mean, and geometric mean. They may not always be the parameter of interest, but in most cases, provide valuable information even if they are not. The **mode** is the value that occurs most often in the set of observations. The **median** is the middle value of a set of observations after ranking in ascending or descending order.

The **arithmetic mean** is a more technical name for what is known as the mean or average. It has appealing statistical properties and is commonly modeled or explained in statistical analyses in various forms. The **geometric mean** is the mean or average of a set of data, but its values are measured on the logarithmic scale. The geometric mean is best when the logarithms of the observations are distributed normally. The geometric mean dampens the effect of extreme values. The geometric mean is always smaller than the corresponding arithmetic mean. It is not affected by extreme values as the arithmetic mean is.

A normal distribution of observations is a graph where most of the observations lie at or near the middle. As you move away from the middle in either direction, the frequency of observations decreases uniformly, as in Figure 2.2. In practice, the researcher will not see an exact normal distribution but rather an approximation or a smoothing to such.

In summary, the arithmetic mean uses all the data to provide an average value and has good statistical properties but is sensitive to outliers (observations that lie an unusual distance from the other observations). The geometric mean provides the logarithmic average and is not sensitive to outliers. In contrast, the median provides a central value, such that 50% of the observations are smaller and 50% are larger than the median.

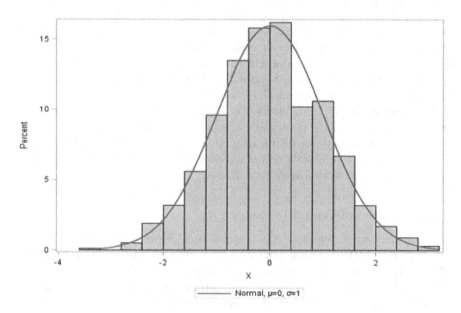

FIGURE 2.2
Example of data that follow a normal distribution

Proportion is the mean of a binary variable (two possible outcomes). As such a proportion is a measure of central tendency. For example, one might use a proportion to describe what fraction of clinic patients tested positive for cancer. A proportion is expressed as a decimal, a fraction, or a percentage. Incidence proportion is the proportion of an initially disease-free population that develops disease, becomes injured, or dies during a specified limited period.

Some related measures:

- **Rate** is a measure of the frequency with which an event occurs in a defined population over a specified period. The incidence rate is the number of new cases per population at risk in a given time period;

- **Prevalence**, or prevalence rate, is the proportion of persons in a population who have a particular disease or attribute at a specified point in time or over a specified period;

- **Prevalence versus incidence**: Prevalence differs from incidence in that prevalence includes all cases, both new and pre-existing in the population at the specified time. Incidence is limited to new cases only. Prevalence is not a rate, so it is not defined by a time interval. It may be defined as the number of cases of a disease that exist in a defined population at a specified point in time.

2.3.3 Measures of Spread or Variation or Dispersion

A measure of spread describes an overall measure of how the values vary from the peak (center) of the distribution. Some measures of spread are the range, interquartile range, standard deviation, and variance. The **range** is the difference between its largest (maximum) value and its smallest (minimum) value. The **interquartile range** (IQR) is a measure of spread used most commonly with the median. It is the central portion of the distribution, representing the difference between the 25th percentile value and the 75th percentile value.

The **standard deviation** describes variability in a set of observations and is used to measure the deviation from the mean. In contrast, the **standard error** describes the variability in a set of values for a particular statistic (a statistic is any value computed from the sample). It relates to differences among a particular statistic. That statistic can be the mean, or the standard deviation, or skewness, or kurtosis as examples. In fact, as an example, one can ask the standard error of the mean, or the standard error of the standard deviation, or the standard deviation of the variance.

Imagine you took several random samples of a certain size repeatedly from a population and obtained the mean and standard deviation for each. The mean and standard deviation are called statistics. The standard deviation of a statistic is called its STANDARD ERROR. In practice, however, one usually has only one sample of observations and can't get the standard deviation of several values for certain statistics. One does not have repeated sampling in practice, yet the standard error is needed. One usually estimates the standard error of a statistic by taking the standard deviation for the statistic based on the sample divided by the square root of the sample size. It is one of the most challenging concepts of descriptive statistics for beginning students to comprehend.

2.3.4 Measures of Relative Standing

2.3.4.1 Single Values

A measure of relative standing tells where an observation lies in comparison to other values. There are percentiles, deciles, quartiles, median, z-value or t-value to name a few. **Percentiles** are used to understand and interpret data. They indicate the values below which a certain percentage of the observations in a dataset is found. Consider ranking the observations in a numerical dataset and splitting them into 100 parts (called percentiles), or 10 parts (called deciles), or 4 parts (called quartiles), or two parts (called median). The 90th percentile is a single value, which has 90% of the observations at or below it. The 4th decile is a single value, which has 40% of the observations at or below it. The 3rd quartile is a single value, which has 75% of the observations at or below it. The median, the halfway point of the distribution – or the 2nd quartile, or 5th decile or 50th percentile – has 50% of observations at or below it. Percentile, deciles, quartiles, and median are represented by a numerical value.

A **z-value** is the number of standard deviations that an observation is located away from the mean. When an observation has a z-value larger than 1.96 in magnitude, one tends to say that relative to the other observations, it is an outlier. Hence, it is common to find that researchers will point to the number 2 (1.96 rounded) since it represents the 97.5th percentile of the standard normal distribution.

A **t-value** is similar in construction to a z-value, except it is best used when the number of observations under consideration is small, and one does not know the population standard deviation from which the sample came. However, it also talks about the number of standard deviations estimated from the mean. While there is one z-distribution (i.e., the normal distribution with mean zero and variance one), there are several t-distributions. Each t-distribution depends on the degrees of freedom (an estimate of the number of independent pieces of information that went into calculating the estimate). As the degrees of freedom increase (depends on the sample size), the t-distribution begins to look more and more like the z-distribution, as seen in Figure 2.3. The degrees of freedom increase as the sample size increases.

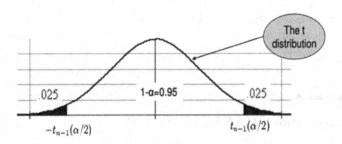

FIGURE 2.3
t-distribution

2.3.4.2 Interval Values

Confidence interval: A confidence interval is a range of possible statistical values (reminder: a statistical value is obtained based on the sample values). It is constructed based on a certain percentage for which the parameter value may lie. Therefore, if one declares that they had a 95% confidence interval of [24, 32] for the population mean of BMI, then one is saying that if many repeated samples of the same size were taken, it is expected to have 95% of these intervals covering the true value. It tells where one *expects* the population values to be standing. For simplicity, it is best to declare that there is a range of possible values for the parameter under study.

2.3.5 Measure of Skewness

Skewness is a measure of the symmetry among the observations in a distribution. It provides the relative size of the two tails in any distribution. Thus, it describes as a measure of symmetry or lack of symmetry among the observations. The rule of thumb when thinking about skewness is between −0.5 and 0.5, the data are fairly symmetrical; if the skewness is between −1 and − 0.5 or between 0.5 and 1, the data are moderately skewed; if the skewness is less than −1 or greater than 1, the data are highly skewed.

2.3.6 Measure of Kurtosis

Kurtosis is a measure of a combined group of observations in the tails relative to the distribution. It is reported relative to the normal distribution, which has a kurtosis of 3. A large positive kurtosis indicates that the data are greatly concentrated near the mean and decline rapidly from the center with heavy tails on both sides of the mean. If the kurtosis is greater than 3, then the data have heavier tails than a normal distribution (more in the tails).

If the kurtosis is less than 3, then the dataset has lighter tails than a normal distribution (less in the tails).

2.4 Results from NHANES Data on BMI

From the NHANES data, we concentrate on information pertaining to BMI measured on a continuum. These data are used to demonstrate some of these measures. There are 6795 observations. The mean BMI is 28.32 with a standard deviation of 6.95. A measure of skewness is 1.23, and kurtosis is 3.42. The values range from 13.18 to 84.87. A 95% confidence interval for the mean BMI is [28.16, 28.49]. The closeness between the mean and median suggests that the distribution may be symmetric. BMI appears to be normally distributed. Statistical packages SAS and R are used to analyze data and, from the outputs, present results and interpretation.

2.4.1 Analysis Results with SAS Program

The following SAS code provides the necessary information for data analysis based on the topics covered in this chapter.

```
proc means data=nhanes mean var Median Mode Skewness
Kurtosis ;
var bmi; run;
proc ttest data=nhanes;
var bmi; run;
```

Measures of relative standing:

```
proc univariate data=nhanes plot normal;
var bmi; run;
```

Here is the SAS output from these SAS procedures.

The MEANS Procedure

Analysis Variable : bmi; run;					
Mean	Variance	Median	Mode	Skewness	Kurtosis
28.322	48.304	27.290	30.080	1.230	3.423

The T-Test Procedure

N	Mean	Std Dev	Std Err	Minimum	Maximum
6795	28.3217	6.9501	0.0843	13.1800	84.870

The mean, variance, median, and mode for BMI are given. The standard error is [6.950/square root (6795)] = 0.084. The measure of skewness = 1.230 and measure of kurtosis = 3.423.

Mean	95% CL Mean		Std Dev	95% CL Std Dev	
28.3217	28.1565	28.487	6.950	6.835	7.069
DF	t Value	Pr > |t|			
6794	335.91	<.0001			

The 95% confidence interval for the mean BMI is [28.157, 28.487]. It suggests that the mean BMI lies between 28.157 and 28.487. The 95% confidence interval for the standard deviation BMI [6.835, 7.069]. It suggests that the population standard deviation lies between 6.835 and 7.069. Also, t-value = 28.322/0.084 = 335.91 tells the number of standard deviations from the mean of zero that the population mean lies.

2.4.2 Analysis Results with R Program

The following R code is used to produce the summary:

```
## Load the R Packages
library(readxl)
library(psych)
## Read in the data from the excel file from "Your
        datafile path"
Nhgh = read_excel("Your datafile path/newnhgh.xls")
attach(nhgh)
## Get the data summary
describe(BMI,na.rm=TRUE)

##   vars  n   mean  sd   median trimmed mad  min   max   range skew kurtosis se
##   BMI 6795 28.32 6.95  27.29  27.69  6.17 13.18 84.87 71.69 1.23 3.42    0.08
```

The mean, variance, median, and mode for BMI are given. The standard error is [6.95/square root (6795)] = 0.08. The measure of skewness =1.230 and measure of kurtosis = 3.423.

2.5 Measures of Association of Two Continuous Variables

The association between two continuous variables is often of interest in data analysis. There are several measures of association between variables. These measures are often in the form of a coefficient. They provide information on how the change in one variable is associated with the change in the other variable. Their relations are either positive, negative, or zero. They usually lie between [−1, 1], although there are some less used measures that do not. A perfect positive relation is +1, and a perfect negative relation is −1. A positive value signifies a positive association with the two variables moving in the same direction, while a negative association signifies the two variables are moving in the opposite direction. A value of zero signifies no association.

One such measure of association is the **Pearson correlation** if one assumes that the two variables are normally distributed, or the **Kendall tau** and **Spearman's rho** if one is not willing to make assumptions about the distribution of the variable. These statistics are usually accompanied by p-values (i.e., probability that given the data, the data support no correlation), which conveys whether the association is significant (small p-values indicate significance). A positive association (see Figure 2.4) shows a line sloping upwards. A value of zero implies no association and is graphically seen as a horizontal line or a line that resembles one. A negative association implies a line sloping downwards.

Pearson's correlation coefficient, Spearman's correlation coefficient, and Kendall's correlation coefficients are the most commonly used measures of association. The latter two methods are usually referred to as nonparametric and suggested for non-normally distributed observations. Nonetheless, because of its known sensitivity to outliers, Pearson's correlation leads to a less powerful statistical test for distributions with extreme skewness or

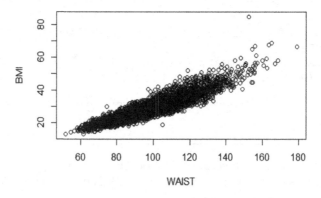

FIGURE 2.4
BMI versus waist

excess of kurtosis (where the datasets with outliers are more likely). In addition, Shong-Chok (2010) found that Pearson's correlation coefficient could have significant advantages for continuous non-normal data, which does not have obvious outliers. Thus, the shape of the distribution should not be the sole reason for not using the Pearson (product-moment) correlation coefficient.

2.6 NHANES Data on Measures of Association for Continuous Variables

The Pearson correlation for BMI and waist had a value of 0.914 with p-value = 0.000. While Kendall tau_b and Spearman's rho were 0.751 and 0.917, respectively. They both have a p-value of 0.000. All three measures of the association gave significant results.

2.6.1 Analysis of Data with SAS Program

```
proc corr data=nhanes;
var BMI Waist;
run;
```

Here is the SAS output from these SAS procedures.

Pearson Correlation Coefficients		
Prob > |r| under H0: Rho = 0		
Number of Observations		
	bmi	waist
BMI		
BMI	1.00000	0.9138
		<.0001
	6795	6556
Waist	0.91380	1.00000
Waist	<.0001	
	6556	6556

There are 6795 observations. The correlation = 0.9138 with p-value <.0001. We conclude that there is a statistically significant correlation between waist size and BMI.

2.6.2 Analysis of Data with R Program

```
# load the library
library(psych)
# combine the data to calculate the correlation
X = cbind(BMI,waist)
# call "corr.test" to calculate the correlation and
        perform statistical test
corr.test(x)

Call:corr.test(x = x)
Correlation matrix
      BMI waist
BMI   1.00 0.91
waist 0.91 1.00
```

> The correlation = 0.91 with p-value <.0001 is less. Conclusion: there is a statistically significant correlation between waist size and BMI.

```
## Sample Size
          BMI    waist
BMI      6795   6556
waist    6556   6556
## Probability values (Entries above the diagonal are
                  adjusted for multiple tests) rounded to 0
##        BMI    waist
  BMI    0      0
  waist  0      0
```

2.7 Association Between Two Categorical Variables

A measure of association between two categorical variables is best seen through a contingency table. A contingency table presents the frequencies of the cross-classification of the categories of the variables.

2.7.1 Contingency Table

There are several measures of association for two categorical variables. Such measures are often a function of a chi-square statistic (a measure of the

TABLE 2.2

A Contingency Table of Gender by Obesity

Gender	Obesity		
	Not Obese	Obese	Total
Female	2155	1268	3423
Male	2289	1083	3372
Total	4444	2351	6795

difference between the observed data and the data expected under the null hypothesis of no association). The measures are accompanied by p-values, which are themselves measures of the strength of the association between the two variables. These measures provide information on how the categories of one variable are associated with the categories of the other. While their relations are either positive or negative, the test statistics are often non-directional.

Table 2.2 represents a case of a categorical (two outcomes; binary) variable cross-classified with another categorical variable (binary). This so-called contingency table consists of cells, which at times can be zero, small, or large. In Table 2.2, there are 3423 females. There are 2155 females who are not obese.

If a cell is a count of zero and would always be zero regardless of how many observations were taken, then it is referred to as a structural zero. If the cell can increase from zero, then the original zero is called a random zero. Data with structural zeros require special methods of analysis (Bishop, Fienberg, and Holland 1974).

2.7.2 Odds Ratio

An odds ratio (OR) represents the odds that a specified outcome will occur when experimental units are exposed to some factor of interest – as compared to the odds that the outcome will occur in experimental units not exposed to the factor. As such, an OR may be thought of as a measure of association between exposure and outcome. It ranges from 0 to infinity. An OR less than 1 means that exposure is associated with lower odds of the outcome. An OR equal to 1 means that exposure does not affect the odds of the outcome. An OR greater than 1 means that exposure is associated with higher odds of the outcome. The distribution of odds ratios is typically skewed to the right.

The OR expresses how much higher the odds of an event occurring among individuals in a group (say A) are as compared to the odds of the event occurring among individuals in a group (say B). In fact, the OR is the ratio of the odds. The odds is the probability that an event will occur compared to the probability that the event will not occur. It gives an indication of how likely an event will occur as opposed to not.

One uses the odds ratio to present a class of models referred to as logistic regression models which are discussed in Chapter 8. It is of great importance to know that the OR is useful with retrospective, prospective, or cross-sectional data.

2.7.3 An Example in Odds Ratio

One hundred patients with a disease received a certain drug for 90 days, resulting in 80 of the cases being improved. At the same time, 100 other patients received a placebo for 90 days, resulting in 40 cases being improved. For ease of computation, one presents the data in a 2 by 2 contingency table, as shown in Table 2.3.

ODDS: The odds of healing in the drug group are $\dfrac{80}{20} = 4$, and the odds of healing in the placebo group are $\dfrac{40}{60} = \dfrac{2}{3} = 0.67$.

ODDS RATIO: Odds ratio is $\dfrac{4}{0.67} = 6$. If one wrote $\dfrac{80 \times 60}{40 \times 20} = 6$, then one can see how the form of the odds ratio has earned the name, cross-product ratio. Thus, patients in the drug group are 6 times more likely to be improved (in odds scale). The odds ratio is greater than one. Hence, one states that there is a positive association between taking the drug and the outcome: improve (row1 with col1). Some may say that the drugs increased effectiveness by 600 times. Consider a rearrangement of the cells in Table 2.4:

ODDS RATIO: Odds ratio is $\dfrac{20/80}{60/40} = \dfrac{1}{6} = 0.167$. Thus, patients in the drug group are 0.167 times more likely to not improve. The odds ratio is less than one. Hence, one states that there is a negative association between taking the drug and not improving.

2.7.4 Relative Risk

A risk ratio (RR), also called relative risk, compares the risk of an event (disease, injury, risk factor, or death) among one group with the risk among another group. The two groups are usually different based on demographic factors or certain risk factors. Often, the group of primary interest is referred

TABLE 2.3

Contingency Table of Group by Outcome

Group	Outcome	
	Healed	Unhealed
Drug	80	20
Placebo	40	60

TABLE 2.4

Contingency Table of Group by Outcome

	Outcome	
Group	Unhealed	Healed
Drug	20	80
Placebo	60	40

to as the exposed group, and the comparison group is referred to as the unexposed group. A risk ratio of 1.0 indicates equal risk in both groups. A risk ratio greater than 1.0 indicates an increased risk for the group in the numerator. A risk ratio less than 1.0 indicates a decreased risk for the group in the numerator. Let us refer to Table 2.3.

PROBABILITY: The probability of healing among patients taking the drug is $\frac{80}{100} = 0.80$. The probability of improving among patients taking the placebo is $\frac{40}{100} = 0.40$. Thus, the ratio of the probabilities is $\frac{0.80}{0.40} = 2.00$. Thus, the drug increases the risk of healing by 100%. One was able to compute the risk, as it is possible to compute the probabilities. This follows as the denominator was fixed (determined before conducting the experiment). Otherwise, one would not be able to compute the relative risk, although one can always compute the odds ratio. Relative risk is the ratio of the probability of the outcome of improving if exposed to drug compared to the probability of the outcome of healing if not exposed to the drug.

2.7.5 Properties of a Diagnostic Test

When a diagnostic test is performed in a patient, the result may lead the diagnostician to declare that the patient has a certain disease. However the patient may not have the disease. Thus the test may be truly positive (disease present) or the test may be truly negative (disease not present). For example, 100 participants who were known to have cancer were tested for cancer. At the same time, 200 patients who do not have cancer were also tested.

Let C be the event that the patient has cancer, *and let* C^- be the event that the patient does not have cancer. Let T^+ be the event that the test is positive and T^- be the event that the test is negative. There are also false positives and false negatives. Consider the results in (Table 2.5).

We note from Table 2.5 that the diagnostic test correctly identifies 95 of the 100 patients with cancer as having cancer, and incorrectly identifies 5 of 100 patients as not having cancer (called false negatives). Similarly, the diagnostic test correctly identifies 180 of the 200 patients without cancer as not having cancer, and incorrectly identifies 20 of the 200 patients without cancer as having cancer (false positives).

TABLE 2.5

Cross Classification of Disease versus the Test result

	Positive (T+)	Negative (T-)
Cancer (C+)	95	5
No Cancer (C-)	20	180

This problem can be seen in other settings. For example, individuals going through the security screening at an airport sometimes have a false detection. Such is the case of a test for certain diseases; there are false positives and false negatives. No one wants a test with a high percentage of false positives or high percent of false negatives. In fact, the desire is to have high sensitivity and high specificity. This is described as follows:

1) **Sensitivity** $(P(T^+ \mid C^+))$ is a measure of probability for a test to correctly identify the presence of an event in an individual who truly has the event. One wishes high sensitivity, where it is measured by $P(T^+ \mid C^+)$, the sensitivity is $\frac{95}{100} = 0.95$ in our example.

2) **Specificity** $(P(T^- \mid C^-))$ is a measure of probability, which is for a test to correctly identify a nonevent in an individual who truly does not have the event. One wishes high specificity, where it is measured by $(P(T^- \mid C^-)$, the specificity is $\frac{180}{200} = 0.90$ in our example.

   In research, one needs to do a balancing act, as at times, there are trade-offs in terms of sensitivity and specificity. Sometimes a test may have very high sensitivity but results in low specificity. The desire is to have both high sensitivity and high specificity.

3) **False positive rate**: A false positive occurs when the test reports a positive result (T+) for a patient who does not have the disease (C-) and the false-positive rate is then calculated as $P(T^+ \mid C^-)$. In our example, there were 20 false positives, and the false-positive rate is $P(T^+ \mid C^-) = \frac{20}{200} = 0.10$.

4) **False negative rate**: A false negative is an event which occurs when the test reports a negative result (T-) for a patient who actually has the disease (C+) and the false-negative rate is then calculated as $P(T^- \mid C^+)$. In our example, there were 5 false-negative events and the false-negative rate is $P(T^- \mid C^+) = \frac{5}{100} = 0.05$.

In practice, once the test is established, the question of interest becomes, if one tested positive (T+), what is the probability that he/she has the event (C+).

This is the predictive positive probability, $P(C^+|T^+)$ as well as the prevalence rate and is defined as:

$$P(C^+|T^+) = \frac{P(T^+|C^+)P(C^+)}{P(T^+|C^+)P(C^+) + P(T^+|C^-)P(C^-)}$$

where $P(C^+)$ is the probability of the event in the population.

2.8 Measure of Belief

In hypothesis testing, one has two hypotheses, null-H_0 and an alternative-H_1 (or research hypothesis). When one tests hypotheses, one needs a test statistic whose distribution is known under the null hypothesis (assuming that H_0 is true) and specification of the magnitude (usually denoted by α) of the allowable error (Type I or False Positive) in making the wrong decision (concluding that H_0 is not true when it is). Then one uses α, and the distribution of the statistic under H_0, to determine the critical region for the test.

For example, when testing a null statement about the unknown mean (μ) of a population, it is reasonable to base inference on the mean of a random sample $\bar{x}$ from the population. One can show that the distribution of the sample mean has mean the same as the population mean (μ) and variance the population variance (σ^2) divided by the size of the sample (n).

$$\bar{X} \sim N(\mu, \sigma_x^2)$$

The mean of the distribution of the sample mean being the same as the population mean tells us that the sample mean is unbiased for estimating μ. A famous theorem (Central Limit theorem) in probability and statistics is that for moderate to large-sized samples from the population, the distribution of the sample mean is approximated well by a **normal distribution** with mean μ and variance σ^2/n.

Now we can show that the sample mean may be **standardized** to produce a Z variable whose distribution has mean 0 and variance 1. To standardize a random variable (in this case the sample mean), subtract its mean and divide by the square root of its variance (called the standard error of the mean). This gives $Z = (\bar{X} - \mu)/(\sigma/n^{1/2})$, and can be used as a test statistic if σ is known (the value of μ is stated under the null hypothesis, n is specified in advance of sampling, and $\bar{X}$ can be computed from the random sample). If σ is unknown, its estimate (symbolized as s) may be computed from the sample, but when s replaces σ in the above equation, Z becomes T, which has a t-distribution with $n - 1$ degrees of freedom (df), i.e., $T = (\bar{X} - \mu)/(s/n^{1/2})$.

Testing at the $\alpha = 0.05$ level, and assuming that H_0 is rejected for values of the test statistic that are too large (positively or negatively), the critical region becomes numbers whose absolute values exceed 1.96 (the 97.5th percentile from the normal zero/ one ($N(0,1)$] distribution). It is noted that as the degrees of freedom become large, the t-distribution approaches the $N(0,1)$ distribution (e.g., when df = 1000, the 97.5th percentile from the t-distribution is 1.962). In summary, the steps in hypothesis testing are:

1. Specify the hypotheses: H_0: $\mu = \mu_0$ versus H_1: $\mu \neq \mu_0$ (called a 2-sided alternative);
2. Select the test statistic, say $Z = (\bar{X} - \mu)/(\sigma/n^{1/2})$;
3. Select the False positive rate, say $\alpha = 0.05$;
4. Decide on the size of the sample, n;
5. Determine the Critical Region (CR), say $(-\infty. - 1.96)$ or $(1.96, \infty)$;
6. Conduct the experiment and gather the data;
7. Compute the value of the test statistic, say z;
8. Make a decision. Reject H_0 if $|z| > 1/96/$.

Instead of deciding to reject (or not) the null hypothesis based on whether the value of the test statistic lies in the CR, one can compute the p-value for the test, and if the p-value is less than α, then reject the null hypothesis. The p-value is defined as the probability of the data obtained (observed) or data more extreme, given the null hypothesis is true. The p-value is an estimate of the strength of the data regarding H_0. It is a measure of the belief one has based on the test statistic in support for or against H_0. The larger the p-value, the more apt one is to believe that H_0 is true and have no belief in H_1.(see Figure 2.5).

2.9 Analysis of Data on Association Between Two Categorical Variables

We use the NHANES data to investigate the association between obesity (BMI > 29.9 and gender. We use SAS and R to analyze the data.

2.9.1 Analysis of Data with SAS Program

```
*Measures of Association;
data new;
set nhanes;
```

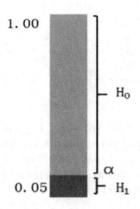

A p-value of 1.00 supports $H0$ which means that $Z \vee T$ is 0. This happens only if the sample mean is identical to the hypothesize value under H0. When the p-value is 0 one can reject H0 and declare $H1$. The question is what happens when $0<p<1$. Think of p-value as providing a degree of belief in the data used as supporting the null hypothesis. Less belief of $H0$, the smaller is the p-value.

FIGURE 2.5
Null and alternative hypothesis and p-value

```
Obese=(bmi>29.9);
run;
proc freq data=new;
tables new_bmi*gh;
run;
proc freq data=new;
tables new_bmi*sex/all;
run;
```

Obese	Gender		
	Female	Male	Total
Not obese	2155	2289	4444
	31.71	33.69	65.40
	48.49	51.51	
	62.96	67.88	
Obese	1268	1083	2351
	18.66	15.94	34.60
	53.93	46.07	
	37.04	32.12	
Total	3423	3372	6795
	50.38	49.62	100.

Statistic	DF	Value	Prob
Chi-square	1	18.2164	<.0001
Likelihood ratio chi-square	1	18.2310	<.0001

Continuity adj. chi-square	1	17.9993	<.0001
Mantel-Haenszel chi-square	1	18.2137	<.0001
Phi coefficient		−0.0518	
Contingency coefficient		0.0517	
Cramer's V		−0.0518	

Association of obesity by gender (two categorical variables is measured by the chi-square, likelihood ratio chi-square, etc. signifies a statistically significant association. The Prob (p-value) is < 0.0001, which indicates that the association is statistically significant. Phi-coefficient, contingency, and Cramer's V are measures of association.

Fisher's Exact Test	
Cell (1,1) Frequency (F)	2155
Left-sided Pr <= F	<.0001
Right-sided Pr >= F	1.0000
Table Probability (P)	<.0001
Two-sided Pr <= P	<.0001

Fisher's exact test is applicable for all sizes of data. Assume that the observations are independent, and we are sampling without replacement. A p-value of <.0001 suggests a statistically significant association.

2.9.2 Analysis of Data with R Program

```
## Load the R package
library(gmodels)
## Call R function "CrossTable" with bmi and gh ##
freq1 = CrossTable(nhgh$new_bmi,nhgh$gh,expected = TRUE,
      prop.r=TRUE,
      prop.c=TRUE,prop.t=TRUE, prop.chisq=TRUE, chisq =
            TRUE,
      resid = TRUE,sresid=TRUE, asresid=TRUE)
## Total Observations in Table: 6795
## Pearson's Chi-squared test
## Chi^2 = 665.8272   d.f. = 98   p = 2.902884e-85
```

```
Freq = table(nhgh$new_bmi,nhgh$gh) ## bmi&gh
Perc = prop.table(freq)*100
freq2= CrossTable(nhgh$new_bmi,nhgh$sex,expected = TRUE,
prop.r=TRUE,
        prop.c=TRUE, prop.t=TRUE, prop.chisq=TRUE, chisq
            = TRUE,
        fisher = TRUE,resid = TRUE, mcnemar=TRUE,
        Sresid=TRUE, asresid=TRUE) ##bmi&sex
```

```
## Total Observations in Table:     6795
## nhgh$new_bmi |    female |      male |Row Total |
## ------------ |--------- |----------- -|--------- |
##           0|     2155 |      2289 |     4444 |
##             | 2238.677 |  2205.323 |          |
##             |    3.128 |     3.175 |          |
##             |    0.485 |     0.515 |    0.654 |
##             |    0.630 |     0.679 |          |
##             |    0.317 |     0.337 |          |
## ------------ |--------- |----------- -|--------- |
##           1 |     1268 |      1083 |     2351 |
##             | 1184.323 |  1166.677 |          |
##             |    5.912 |     6.002 |          |
##             |    0.539 |     0.461 |    0.346 |
##             |    0.370 |     0.321 |          |
##             |    0.187 |     0.159 |          |
##------------ |--------- |----------- -|--------- |
## Column Total |     3423 |      3372 |     6795 |
##             |    0.504 |     0.496 |          |
## Statistics for All Table Factors
## Pearson's Chi-squared test
## Chi^2 = 18.21638      d.f. = 1     p = 1.971754e-05
## Pearson's Chi-squared test with Yates' continuity
correction
## Chi^2 = 17.99934      d.f. = 1     p = 2.20982e-05
## McNemar's Chi-squared test
## Chi^2 = 293.0675      d.f. = 1     p = 1.067072e-65
## McNemar's Chi-squared test with continuity correction
## Chi^2 = 292.4937      d.f. =1      p = 1.423033e-65
```

The association of obesity by gender (two categorical variables) as measured by the chi-square, likelihood ratio chi-square, etc. signifies a statistically significant association. The Prob (p-value) is < 0.0001, which indicates that the association is significant.

```
X = c(2155,2289,1268,1083)
Table = matrix(x,nrow=2)
fisher.test(table,alternative ="two.sided")
## Fisher's Exact Test for Count Data
### data: table
## p-value = 2.047e-05
```

> Fisher's exact test, appropriate for all sizes of data, has a p-value that indicates the association is statistically significant.

```
## alternative hypothesis: true odds ratio is not equal
to 1
## 95 percent confidence interval:
##0.7264999 0.8899704
## sample estimates:
## odds ratio
##   0.804132
```

> The odds ratio is 0.804 with a confidence interval [0.726, 0.890], which is statistically significantly different from 1. This suggests a negative association. According to the table, this means that the first row is associated with the second column. Therefore, females are more likely to be obese.

Results: There is a positive association between gender and obesity. This was based on a Pearson chi-square value of 18.216 with a likelihood ratio of 18.23 [p-value = 0.0001]. A test appropriate for any sample size, Fisher's exact test, gave p-values near 0 (0.000). The odds ratio is = (1269*1083)/(2289*2155) = 0.804. The 95% confidence interval (CI) for the odds ratio is [0.726, 0.890] (see Wasserstein and Lazar 2015). Our practice is to also provide a 95% CI.

Summary

Question: This chapter dealt with descriptive measures and some two-variables-at-a-time analysis. One uses a statistic either to estimate the population parameter of interest or to test some research statement about the parameter. In so doing, one needs to account for the fact that the sample is for a subset of the population and not the population. For example, is the mean BMI different from 27.5? This is hypothesis testing rather than estimation.

Answer: One would need to construct a test statistic or obtain a confidence interval to tell if the mean is different from 27.5. The test statistic can be parametric or nonparametric. We point out the differences in these methods in Chapter 3.

2.10 Exercises

Examine the NHANES data for the variable "weight":

1. Obtain the measures of central tendency, mean, mode, median;
2. Obtain the measures of dispersion, range, variance, standard deviation;
3. Obtain the measures of relative standing, 95% percentile, 3rd decile, 1st quartile, and z-value for the 5th percentile;
4. Does weight seem to be normally distributed?
5. What is the association between age and weight?
6. What is the measure of association between treatment and gender?

Write a summary of your findings in such a way that someone who knows no statistical jargon can understand,

References

Bishop, Y.M., Fienberg, S.E., Holland, P.W.: *Discrete Multivariate Analysis Theory and Practice*. Springer, New York (1974)

Shong Chok, N.: *Pearson's versus Spearman's and Kendall's Correlation Coefficients for Continuous Data*. University of Pittsburg (2010)

3

Statistical Modeling of the Mean of Continuous and Binary Outcomes

3.1 Research Interest/Question

In this chapter, we present the statistical models for testing the mean of continuous responses and of binary responses. In particular, tests for the mean of a normal (continuous) random response and for the mean of a binomial (binary) random response measure are presented. We apply these statistical models to the German Breast Cancer Study data as examples. In particular, modeling the mean number of cancerous nodes based on a random sample of data from the distribution is illustrated. As such, in statistical form, the null hypothesis $H_0 : \mu = 2.0$ in comparison to the research hypothesis (alternative) $H_1 : \mu > 2$, where μ is the unknown mean of the distribution of nodes. In the first part of the chapter, there is no interest in a predictor (also referred to as driver, or a covariate). Interest is in the mean of the distribution of nodes, and whether the proportion of patients with tamoxifen, p, is less than 50%. This hypothesis is expressed in statistical form as the null hypothesis $H_0 : p = 0.5$ and the research hypothesis as: $H_1 : p < 0.5$.

3.2 German Breast Cancer Study Data

In 1984, the German Breast Cancer Study Group recruited 686 patients with primary node-positive breast cancer into the Comprehensive Cohort Study (Schmoor et al. 1996). Both randomized and nonrandomized patients were eligible. About two-thirds were entered into the randomized portion to examine the effectiveness of three versus six cycles of chemotherapy as well as hormonal treatment with tamoxifen. Follow-up at about five years resulted in 312 patients who had at least one recurrence of the disease or who had died. We analyze these data from the German Breast Cancer Study to demonstrate statistical testing for the mean of continuous outcome measures and mean for the binary outcome measures (proportion binary response) (Sauerbrei and Royston 1999). A subset of the data is shown in Table 3.1, and a frequency distribution for the nodes is shown in Figure 3.1. Table 3.1 consists

DOI: 10.1201/9781003315674-3

TABLE 3.1

Subset of German Breast Cancer

Id	Age	Meno	Size	Nodes	Hormon
132	49	premenopausal	18	2	no tamoxifen
1575	55	Postmenopausal	20	16	no tamoxifen
1140	56	Postmenopausal	40	3	no tamoxifen
769	45	premenopausal	25	1	no tamoxifen
130	65	Postmenopausal	30	5	had tamoxifen
1642	48	premenopausal	52	11	no tamoxifen
475	48	premenopausal	21	8	no tamoxifen
973	37	premenopausal	20	9	had tamoxifen
569	67	Postmenopausal	20	1	had tamoxifen
1180	45	premenopausal	30	1	no tamoxifen

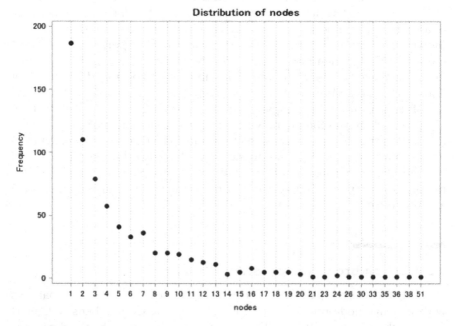

FIGURE 3.1
Graph of distribution of nodes

of ID, age, menopause (meno), size, number of nodes, and whether or not they had tamoxifen.

It appears from the distribution of the number of nodes in Figure 3.1 that most patients had less than two nodes. Our interest lies in the mean number of nodes. It is not the individual sizes of the nodes that is our interest but the mean size of the population of nodes. These observations are skewed to the right. By skewed to the right, we mean that most of the observations lie near or left of the median, and fewer and fewer observations lie to the right of center. In analyzing these data, we make mention of two kinds of tests, parametric and nonparametric. These tests are given for testing measures of central tendency (mean and median) (see Section 3.3) and for testing proportion (mean of a binary variable) (see Section 3.4).

3.3 Parametric Versus Nonparametric Tests

A parametric statistical test for testing hypotheses is constructed based on certain conditions about the population from which the random sample for the research was drawn. In the present case, the assumptions for conducting these tests require that the observations be independent. By independent, we mean that the chance of including an observation in the sample is not affected by nor does it affect the chance that any other observation is included in the sample. The observations are measured on an interval scale and come from normally distributed populations.

A nonparametric statistical test is a test where the model does not rely on any specific distributions about the population from which the sample was drawn. Most nonparametric tests are used for data on an ordinal scale, but some apply to data on a nominal scale. Nonparametric tests rely on the order or ranking of the observations. The given data are changed from scores to ranks or signs. A parametric test focuses on the mean of the population, but the corresponding nonparametric test focuses on the median (Armitage and Berry 2008).

There are two main reasons for performing a statistical analysis whether with parametric test or nonparametric test:

1. Estimation of the population parameter; and
2. Tests of hypotheses regarding the parameter of the distribution.

When there are assumed known distributional properties for the variable of interest (response) in the data under analysis, one should choose a parametric test. Parametric tests are more powerful than nonparametric tests. This is not surprising, as more information is available with a parametric test.

3.4 Statistical Model for Continuous Response

3.4.1 Parametric Tests

In statistical analyses, there are two hypotheses, a null hypothesis H_0 and an alternative hypothesis H_1. If the research hypothesis of interest (H_1), is supported, this may lead to changes to the present state of research and practices as it requires the institution of some new policies or procedures. The other hypothesis is the null hypothesis (H_0). It represents the present state of the system. This is referred to at times as the do-nothing hypothesis. In the present research, we want to determine if the mean of the nodes is less than a value of two (Michaud 2008).

Define the hypotheses $H_0 : \mu_{nodes} = 2$ and $H_1 : \mu_{nodes} > 2$. The primary interest is in the mean of the distribution. Of course, as seen with the observations in Figure 3.1, there are observed nodes greater than a value of two and some even a lot larger than a value of two. However, we are looking at the mean as a location parameter. In like manner, one can ask about the mode, the median, and even the variance, or some other parameter of the assumed distribution (Snedecor and Cochran 1989).

As the interest is in the mean parameter, it is helpful if we know from what distribution the outcomes originated. A common approach is to assume that since the outcomes are on a continuous scale, the distribution of the responses, is possibly normal (or Gaussian). In other words, the population of nodes, in this case, is expected to have most observations lying in the middle and declining gradually at the same rate as one moves away from the middle to either side. One may have problems accepting that such is the case, as shown in Figure 3.1. It is good practice to look at the sample of observations and the literature on the topic to guide the selection of the distribution, but that may not always be prudent. However, we are interested in knowing if we took several samples how the mean of these samples will behave. The central limit theorem comes to the aid – especially when we are working with "large" size samples.

One needs to construct a statistical test that uses the key information available (mean, sample size, standard error) to obtain a distance measure between the hypothesized value and mean of the sample values (women in the breast cancer study). It is referred to as the statistical test value. Based on that test statistic value, one makes a decision about the hypothesis. Given the distribution of test statistic value, a method is devised to tell if the value is large enough to decide that there is sufficient evidence to support the null hypothesis, H_0. One wants to be overcautious when making conclusions that lead to changes in the present state of things. Therefore, a requirement for more stringent evidence to support the research hypothesis and to reject the null hypothesis, H_0, is needed. As supporting the H_1 may lead to changes in the present state of nature, one relies on the distribution that one believes reflects the behavior of the test statistic values. Statistical theory tells us, in a situation like this where the population variance is not available to the researcher, to rely on the t-statistic when there is a small sample size and to follow the z-statistic when sample size is large. Whatever distribution one relies on, the p-value (assuming the null hypothesis, the probability of observing the given data or data more extreme) is based on that distribution we used (Greenland et al. 2016). We refer the reader to the ASA statement on p-values (Wasserstein and Lazar 2016) referenced in Chapter 2.

3.4.2 Statistical Model

A test statistic value originates from the statistical model that one chooses to adopt. In the test of the mean, the statistical model in its simplest form is:

$$\text{Observation}_i = Y_i$$

$$= \mu + \text{Error}_i$$

where observation$_i$ (i.e., Y_i) denotes the value obtained from the ith subject sampled and μ is the overall mean that is unknown but hypothesized about, and *Error*$_i$ represents the difference between the ith observation and the overall mean, often called random measurement error. *Error*$_i$ represents information beyond or below the mean that the ith subject may have but is unknown to the researcher. This is referred to as random variation. It is customary to make assumptions about this random variation. In particular, a normal distribution assumption is often made about the *Error*$_i$ as having a mean of 0 and a variance of σ^2. Alternatively, but equivalently one can say that a normal distribution assumption is made about the observation$_i$(Y_i) but with a mean of μ and a variance σ^2. Moreover, we assume that the outcomes are independent in this setting.

3.4.3 Statistical Test – One Sample t-Test

The statistical test is made up as a ratio of the mean difference to the standard error (a measure of the sample standard deviation to square root of the sample size). The test value provides a p-value (area beyond the test statistic value of the distribution of the test statistic) that is used to determine the strength of the statistical test value.

The p-value is based on the distribution of the test statistic under H_0, It provides guidelines for a cut point to decide if the resulting value is small enough to make a decision to support or not to support the research hypothesis. Large test statistic values will put the point further into the tail which gives less area in the tail and leads to smaller p-values. The p-value is the area in the tail of the distribution beyond the test statistic value. We refer to the cut point as the value so that the area under the distribution curve and beyond the value is α (or as the probability that we conclude that something is significant when in fact it is not) referred to as the significance level or the probability of the type 1 error but specified prior to data collection. In short, a rule of thumb is, if the p-value is less than α we increase our degree of belief in H_1 and recommend to reject the null hypothesis H_0. As one becomes more involved in thought process of p-values, one will see some alternate proposition from statisticians.

3.4.4 Alternative Methods to One-Sample t-Test

CONFIDENCE INTERVAL: A confidence interval is a range of possible values of the parameter under measure on an associated level of belief. It can be used to determine the decision in a hypothesis test. The interval has a lower limit or an upper limit (corresponding to a 1-sided H_1) or both (corresponding to a 2-sided H_1). It is constructed based on the distribution of the test statistic, often the normal distribution (most of the means are in the middle, and as you move from the middle, the frequency of observation gradually declines on both sides of the center). If the range is narrow and the limits exclude the hypothesized mean value then one might determine that the hypothesis should be rejected and H_1 supported.

When the distribution based on the information in the sample cannot support the mean as from a normal population, then one may question the use of a parametric test, as the assumption necessary for the one-sample t-test is not satisfied. An alternative approach is to engage in a transformation of the response variable; say, use the log or square root to make it consistent with being normally distributed. Another approach is to use a generalized linear model, which is discussed in Chapter 8 or a nonparametric test, as in Section 3.4.2.

> **Questions:** Do the nodes seem to follow a normal distribution as assumed when using the t-test?
>
> (Normality assumption)? Is this a random sample (Random observations)?
>
> **Answers:** One wishes to determine which of the assumptions are met based on the sampled observations. Normality: One validates this assumption by examining plots and the use of statistical tests. Plots, such as box plots and probability plots are often used. Skewness and kurtosis are also key measures to help in the assessment of normality because of its uniqueness. Tests such as Shapiro-Wilk, Kolmogorov-Smirnov, Cramer-von Mises, or Anderson-Darling tests are useful in checking for normality. If any of these tests reject (i.e., small p-values) the hypothesis of normality, the data should be questioned in their support of the normality assumption, as shown in Table 3.2. A transformation of the data to achieve normality is also possible. Common transformations such as square-root or logarithm are used to normalize the response variable in the data. Of course, the transformed variable is subject to normality checks, as in (Table 3.2).
>
> **Random Sample:** The validity of this assumption depends on the method used to obtain the sample observations. If the method used

TABLE 3.2

Tests for Normality with Small p-values Suggesting Non-normal

Test	Statistic			p Value
Shapiro-Wilk	W	0.710479	Pr < W	<0.0001
Kolmogorov-Smirnov	D	0.231964	Pr > D	<0.0100
Cramer-von Mises	W-Sq	9.253377	Pr > W-Sq	<0.0050
Anderson-Darling	A-Sq	51.15448	Pr > A-Sq	<0.0050

ensures that each individual in the population of interest has a probability of being selected for this sample, one has a probability random sample. If the observations are obtained on the basis of happenstance or convenience or some other non-probability method, then the test is questionable.

3.5 Analysis of the Data

A demonstration of these statistical methods as discussed is used to investigate if research hypothesis $H_1 : \mu > 2$ on the German Breast Cancer study data using two statistical programs: SAS and R are presented. Truncated versions of the outputs are provided for ease in presentation. At times, comments, interpretation, findings, and questions are made to guide the reader.

3.5.1 Analysis of the Data Using SAS Program

```
proc ttest data=germanbreast;
var nodes;
run;
```

These SAS codes gives this output.

The TTEST Procedure
Variable: nodes (nodes)

N	Mean	Std Dev	Std Err	Minimum	Maximum
686	5.010	5.476	0.209	1.000	51.000

Mean	95% CL Mean		Std Dev	95% CL Std Dev	
5.010	[4.600	5.421]	5.476	[5.200	5.782]

DF	T VALUE	PR > |T|
685	14.40	<.0001

The mean is 5.01 based on 686 observations. The standard deviation is 5.476, but the standard error (standard deviation of the mean is 0.209 = (5476/sq root(686)).

The p-value is <.0001. Since we have a one-sided test, and the distribution is symmetric, we divided the 2-sided p-value by 2 (p-value/2) to obtain the correct p-value for our one-sided test. Since the p-value is so small, the data support the research hypothesis. The 95% CI for the mean [4.60, 5.42] provides us with a range of possible values for the population mean. The 95% CI for the standard deviation [5.20, 5.78] provides us with a range of possible values for the population standard deviation. Since the value falls outside the range [4.6, 5.42], we say the mean is probably greater than two.

3.5.2 Analysis of the Data Using R Program

```
t.test(germanbreast$nodes,mu=2)
## One Sample t-test
## data: germanbreast$nodes
## t = 14.399, df = 685, p-value < 2.2e-16
## alternative hypothesis: true mean is not equal to 2
## 95 percent confidence interval:
## 4.599739 5.420669
## sample estimates:
## mean of x
## 5.010204
```

The p-value is <2.2e-16. We have a one-sided test, so we take the p-value/2 to obtain the correct p-value. Since p-value is so small we conclude that the data support the research hypothesis. The 95% CI for the mean provides us with a range of possible values for the population mean [4.60, 5.42].

Comment: An example where the research question is directional (µ mean is greater than two) is presented. However, one may have a research question with no direction (e.g., mean not equal to the value two). When there is a directional hypothesis and the distribution is symmetric, one must adjust the p-value by taking ½ of the result given in the outputs (taking ½ should not be done if the distribution is not symmetric. In this case it is the t-distribution, so it is symmetric). These statistical programs usually assume one is conducting a two-sided test [≠] by default. Some statisticians believe that one should not necessarily compare the p-value with α but rather interpret the p-value as a measure to tell about the probability of a situation based on the sample of observations used.

3.6 Continuous Response with No Covariate: Nonparametric Test

Consider a test statistic that does not rely on a distributional assumption about the population but only requires independent observations and the ability to rank the observations (Randles and Wolfe 1979). A common test used in such a situation is the **Wilcoxon signed-rank test**. It is the nonparametric equivalent of the one-sample t-test. The Wilcoxon signed-rank test procedure assumes that the sample of observations was randomly obtained from a symmetrical population. This symmetrical assumption does not need to be exactly normal, as long as there seems to be roughly the same number of values above and below the median. It computes a test statistic that is compared to an expected value. The test statistic comprises summing the ranked differences of the deviation of each observation from a hypothesized median beyond the hypothesized value.

Another nonparametric equivalent of the one-sample t-test is the sign test. It has a different set of assumptions as opposed to the Wilcoxon signed-rank test. It assumes that data obtained come from a non-normal distribution. It makes no assumption about the shape of the population distribution. It does not rely on symmetric patterns. The population can be skewed to either the left or the right. The sign test is fit for testing a hypothesized median value from a single population. It also provides confidence intervals. Its p-value is based on a binomial distribution.

3.7 Data Analysis in SAS and R

A demonstration of these nonparametric tests to analyze hypotheses using the German Breast Cancer study data with two statistical programs: SAS and R are given. The output, in whole or in truncated form, is provided for ease of presentation. At times, comments, interpretations, and findings, or even questions are provided to help the reader.

3.7.1 Analysis of the Data Using SAS Program

```
*** Nonparametric;
proc univariate data=Germanbreast normal plot;
var nodes;
run;
```

These SAS codes provide this output.

Variable: nodes (nodes)

MOMENTS			
N	686	Sum Weights	686
Mean	5.01020408	Sum Observations	3437
Std Deviation	5.47548332	Variance	29.9809176
Skewness	2.88475933	Kurtosis	13.3134251
Uncorrected SS	37757	Corrected SS	20536.9286
Coeff Variation	109.286633	Std Error Mean	0.20905492

Tests for Location: Mu0 = 2

Test		Statistic	p Value	
Student's t	t	14.39911	Pr > |t|	<.0001
Sign	M	101	Pr >= |M|	<.0001
Signed Rank	S	58123.5	Pr >= |S|	<.0001

It gives measure of central tendency, measures of spread, measure of skewness, and measure of kurtosis. It reports the student t along with the sign t and the sign-ranked t along with p-values.

The p-value is <.0001 for the nonparametric tests (sign test, signed-rank test); the data provide evidence that the median nodes are greater than 2.

3.7.2 Analysis of the Data Using R Program

```
wilcox.test(germanbreast$nodes,mu=2)
## Wilcoxon signed-rank test with continuity correction
## data: germanbreast$nodes
## V = 141210, p-value < 2.2e-16
## alternative hypothesis: true location is not equal to 2
```

> The p-value is < 2.2e-16 for the Wilcoxon signed-rank test; the data provides evidence that the median nodes are greater than 2.

Comment: Some of the nonparametric tests need the assumption of symmetry. A common way to evaluate symmetry is from the density trace on the histogram. The word "histogram" comes from the Greek *histos*, meaning pole or mast. Hence, the direct definition of "histogram" is "pole chart". A histogram is used to display the distribution of continuous data values. A histogram is a graph of the frequency distribution in which the vertical axis represents the count (frequency or converted to %), and the horizontal axis represents the possible range of the data values. The height of each histogram bar provides a measure of the density of data values within the bar. Thus, the density trace is a smoothed histogram.

3.8 Statistical Models for Categorical Responses

In analyzing categorical response data, let us consider the binary variable Hormon (patient who had tamoxifen versus patient who did not). The data are summarized in Table 3.3 with a graphical presentation in Figure 3.2. There are 35.86% patients with tamoxifen and 64.14% patients had no tamoxifen. The question is whether the population proportion with tamoxifen is less than 50%, $H_1 : p_{tam} < 0.50$?

3.8.1 Parametric Tests

In this section, the case when the response is binary as opposed to the response is continuous (Section 3.4) is considered. For example, a researcher claimed that there are less than 50% of breast cancer patients who had taken tamoxifen. Then, the interest is in the mean of a binary random variable. Consider the hypotheses:

TABLE 3.3

Distribution of Hormon

Hormon	Frequency	%	Cumulative Frequency	%
had tamoxifen	246	35.86	246	35.86
no tamoxifen	440	64.14	686	100.00

Percentage

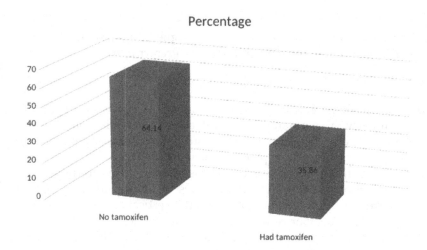

FIGURE 3.2
Percentage bar graphs

$$H_0 : P_{tamoxifen} = 0.50 \, vs. \, H_1 : P_{tamoxifen} < 0.50$$

In the null (no difference) hypothesis form, the proportion of tamoxifen is considered equal to 0.50. The alternative "hypothesis" signifies that tamoxifen exists in less than 50% of the population. The question is still about the mean (proportion or probability).

In the German Breast Cancer data, there are 686 patients. Assume that each patient has the opportunity to take tamoxifen or not. Then this is a Bernoulli trial. The total number of patients taking tamoxifen follows a binomial distribution: $\text{Number}_{tamoxifen} \sim \text{Binomial}[686, P_{tamoxifen}]$, and the patients are assumed independent. Since the outcome is binary, the data may be thought of as consisting of a series of ones (tamoxifen) and zeroes (no tamoxifen). An average of these outcomes is a proportion which lies between [0, 1]. A binomial distribution assumes that you have two possible outcomes, the total number of subjects is known a priori, the outcomes are independent (the mechanism that gives rise to each of the outcomes is not influenced nor does it influence by other outcomes), and the probability of a certain outcome remains the same over the total number of subjects. The fact that the probability remains the same does not mean the outcomes will be the same. For example, flip a coin 686 times; you will not get a head each time though the probability of a head remains the same each time. We refer to this as identical trials.

3.8.2 Test of Proportions [z-statistics]

Consider a test about the proportion of an occurrence! A test of the proportion is derived as follows: consider a measure that takes the key information (proportion, sample size, standard error) from the sample of 686 women into a single value. Our statistical test value is obtained as:

$$Z = \frac{observ.\ proportion - hypo.\ proportion}{Sq\ root\left[\ hypo.\ proportion\left(1 - hypo.proportion\right)/n\right]}$$

where *observ. proportion* is the observed proportion, *hypo. proportion* is the hypothesized proportion, and n is the sample size. It is called a z-test statistic. It provides the basis of a decision about the null hypothesis versus the research hypothesis.

> **Decision:** Once one obtains the statistical test value, one needs a method to tell the researcher when it is large enough to decide that there is sufficient evidence to support the research hypothesis. One relies on the distribution that the test statistic follows a normal distribution. It provides a p-value (i.e., under the null hypothesis, the probability of the data observed or data more extreme) based on the normal distribution. In this case, one relies on the z-distribution. While the p-value is obtained, one needs a cut point to determine when it is small enough to make a decision about significance. It is referred to as the magnitude of the probability of type I error or significance level or α level, which is decided in advance by the researcher or the client. An α of 0.05 is often set before the analysis begins and since p <.0001, (and less than α) there is sufficient evidence to say that the proportion of patients who had tamoxifen is significantly less than 50%, $P_{tamoxifen} < 0.50$.

Demonstrations on the use of these statistics to analyze these hypotheses using the German Breast Cancer study data use two statistical programs: SAS and R. The outputs or their truncated versions are provided. At times, we provide comments, interpret findings, or ask questions and provide answers to help the reader.

3.8.3 Analysis of the Data Using SAS Program

```
data newbreast;
set Germanbreast;
if hormon='no tamoxifen' then newhormon=0;
else if hormon='had tamoxifen' then newhormon=1;
run;
proc catmod data=germanbreast;
      model hormon=;
run;
```

The CATMOD Procedure

Analysis of Maximum Likelihood Estimates				
Parameter	Estimate	Standard Error	Chi-Square	Pr > ChiSq
Intercept	−0.5814	0.0796	53.34	<.0001

> If we were to take $(z=-0.581/0.0796)^2 = 53.34$ = Chi-square statistic. When you square a Z, you get a chi-square with 1-degree of freedom. The p-value must be divided by two as the test is one-sided. Thus, we have sufficient evidence to say that the proportion with tamoxifen is less than 0.50.

3.8.4 Analysis of the Data Using R Program

```
chisq.test(table(germanbreast$newhormon))
## Chi-squared test for given probabilities
## data: table(germanbreast$newhormon)
## X-squared = 54.863, df = 1, p-value = 1.292e-13
binom.test(x=246,n=686,p=0.5)
## Exact binomial test
## data: 246 and 686
## number of successes = 246, number of trials = 686,
                         p-value =1.19e-13
## alternative hypothesis: true probability of success
              is not equal to 0.5
## 95 percent confidence interval:
## 0.3226577 0.3957658
## sample estimates:
## probability of success
##     0.3586006
```

> The value of the test statistic (chi-squared) = 54.863, and one-sided p-value = 1.292e-13/2 (The test is one-sided). Thus, we have sufficient evidence to say that the proportion of patients treated with tamoxifen is less than 50%.

3.9 Exact Tests

The Z-test is considered an asymptotic test, as it is an approximation of the distribution based on large numbers of observations. We now consider the binomial test of proportions to determine an exact test. There is no approximation of distributions. The hypothesis testing construct is:

$$H_0 : p = p_0 \text{ versus } H_1 : p < p_0$$

where p is the probability of a success, $1 - p$ is the probability of a failure, and p_0 is the null hypothesized value of the true proportion of successes. An estimate of p is the # successes/number of trials (number of successes divided by the number of trials), where the trials are independent, and the outcome of each trial is either success or failure. This could cover an experiment as simple as flipping a well-balanced coin n times, recording the number of heads, and testing whether the coin is well-balanced. Then

1. H_0: $p = \frac{1}{2}$ versus H_1: $p < \frac{1}{2}$, where p = the probability of a head in a single flip or trial;
2. H_0: $p = 0.75$ versus H_1: $p > 0.75$ where p is the proportion of operations conducted at a specified hospital that were successful.

Asymptotic test: The tests of these hypotheses can be carried out by using an asymptotic assumption, such as

$$Z = (p^* - p_0) / [var(p^*)]^{1/2},$$

where p^* is proportion of successes out of n trials, and $var(p^*) = p^*(1-p^*)/n$, and determining the p-value, which is the area under the normal zero-one [N(0,1)] distribution to the right of the value of Z. This test is often referred to as the asymptotic test of a proportion. A $100(1-\alpha)\%$ confidence intervals may also be computed, with a

$$lower\ limit = p^* - Z_{\alpha/2}[var(p^*)]^{1/2} and$$

$$upper\ limit = p^* + Z_{\alpha/2}[var(p^*)]^{1/2}.$$

Exact test: An exact test of proportions is performed using the binomial theorem and the null hypothesis to produce the probability distribution of the number of successes out of n trials under the null hypothesis. The probabilities of x successes out of n trials follows the binomial distribution such that

$$P[T = x] = C(n, x) p_0^x (1-p_0)^{n-x}, \quad x = 0, 1, 2, \ldots, n$$

where C(n,r) is the combination of n items taken r at a time, $= n!/r!(n-r)!$, and where the exclamation mark ! means factorial; e.g., $C(4,2) = 4!/2!2! = 4 \times 3 \times 2 \times 1/2 \times 1 \times 2 \times 1 = 24/4 = 6$. The p-value is the sum of terms of the distribution corresponding to the number of successes $\geq$ the numerator of p^* times (the number of observed successes). The p-value can be interpreted as the probability of the data observed or data more extreme.

An alternative exact test of proportions is derived by Clopper and Pearson (1934). We demonstrate the use of the binomial test, Z test, and Clopper and Pearson's exact test in the example.

3.9.1 Analysis of the Data Using SAS Program

Recall the German Breast Cancer study data. Our interest is whether the proportion of patients treated with tamoxifen is less than 50% using two statistical programs: SAS and R. The first part of the output is about the binomial test, the second part of the output is about the Clopper and Pearson's exact procedure, and the last part of the output is about the Z test. The outputs or their truncated versions are given. At times, we provide comments, interpret findings, or even present questions to help the reader.

```
proc freq data=germanbreast;
tables hormon/binomial(p=0.5);
run;
```

	Binomial Proportion
	hormon = had tamoxifen
Proportion	0.3586
ASE	0.0183
95% Lower Conf Limit	0.3227
95% Upper Conf Limit	0.3945
Exact Conf Limits	
95% Lower Conf Limit	0.3227
95% Upper Conf Limit	0.3958
Test of H0: Proportion = 0.5	
ASE under H0	0.0191
Z test	−7.4070
One-sided Pr < Z	<.0001
Two-sided Pr > |Z|	<.0001

The binomial proportion test provides a 95% confidence interval of [0.323, 0.396]. Since the null hypothesized value 0.50 is outside the interval, we reject H_0 and conclude we have sufficient evidence to say that the proportion of patients treated with tamoxifen is less than 0.50.

3.9.2 Analysis of the Data Using R Program

```
binom.test(x=246,n=686,p=0.5)
##
## Exact binomial test
##
## data: 246 and 686
## number of successes = 246, number of trials = 686,
p-value = 1.19e-13
## alternative hypothesis: true probability of success
is not equal to 0.5
## 95 percent confidence interval:
##  0.3226577 0.3957658
## sample estimates:
## probability of success
##       0.3586006
```

The binomial proportion test provides a 95% confidence interval of [0.323, 0.396]. Since 0.50 is outside the interval, we have sufficient evidence to say that the proportion of patients treated with tamoxifen is less than 0.50.

Comments: In this chapter, we entertained hypotheses about the mean or median of a distribution. In like manner, we can ask questions about the variance. Such is the case in quality control, we are particularly interested in the variability of the current process. Or medical practitioners are often faced with varying drug doses. Is the drug dispenser providing precise and accurate values? Is the variance equal to, greater than, or less than some predetermined threshold value? These are all questions that require the variance to be investigated. Tests for the population variance is just as possible to construct.

Example: A large pharmaceutical company produces a certain drug in packages and sells the drug in packs targeted to weigh 20 mg. A quality control manager working for the company was concerned that the variation in the actual weights of the targeted 20-gram packs was larger than acceptable and may do serious harm. That is, she was concerned that some weighed significantly less than 20 mg, and some weighed significantly more than 20 mg. Thus, she was concerned about the variation in the doses. She took a random sample off the factory line. The random sample yielded a sample variance of 2.25 mg². Is there a reason to worry since the acceptable standard deviation is 1.5 mg (or variance = 2.25 mg²? A chi-square test (Snedecor and Cochran 1989) is used to test if the variance of a population is equal to a specified value. This test can be either a two-sided or a one-sided test. Later we will have reason to address the test for a variance as a means of checking for homoscedasticity (same variance across subpopulations), an assumption of the experimental data.

3.10 Summary and Discussion on One-Sample Test

In our example, we tested hypotheses about the mean nodes and obtained a p-value from the statistical programs. Our research hypothesis was $H_1 : \mu_{nodes} > 2$; we obtained a p-value from the statistical software program; we divided the p-value by 2, as $\frac{< .0001}{2}$ (divided by 2, as it is a one-sided test). Thus, we concluded that we had sufficient evidence to say that the mean nodes is significantly greater than two. We should note that the reason for dividing the p-value for a two-sided test to get the p-value for the corresponding one-sided test is that the distribution of the test statistic under the null hypothesis is symmetric about the mean. However, if the distribution is not symmetric, dividing the two-sided p-value by 2 is erroneous, and the p-value will have to be determined directly (using the definition of a p-value, which is the probability under the null hypothesis of the data observed or data more extreme).

Question: When the hypothesis is tested for $H_1 : \mu < 2$, we used a t-test statistic. When the hypothesis is tested for $H_1 : P_{tamoxifen} < 0.50$, we used a z-test statistic. Why do we refer to it sometimes as the z-statistic and other times as the t-statistic?

Answer: We formulate these tests under the assumption of the null hypothesis value. When we deal with continuous data, we draw on the validity of a famous theorem in mathematical probability and statistics called the central limit theorem (CLT). The CLT says that, for a large sample of observations, the distribution of the standardized form of the sample mean (the difference between the sample mean and the null hypothesized mean divided by the standard deviation of the sample mean) is well approximated by the N(0,1) distribution. When we test the proportion $[H_1 : P_{tamoxifen} < 0.50]$, we assume the data consists of a series of independent Bernoulli trials. If we know the variance of the distribution, then the test statistic (the standardized sample mean under H_0) is referred to as a Z statistical test; but if the variance is not known, we have to use the sample variance, and the statistic is referred to a t statistical test. We refer to a t-distribution with n-1 degrees of freedom. But when we have binomial data, the variance is related to the mean. If we know the mean (probability) or its hypothesized value, then the variance is known. As such, we have a z-statistic. In this case, both t and z statistics are formed as the statistic value minus the null hypothesis value divided by the standard deviation for the statistic. The ratio is t when the variance is unknown and z when the variance is known.

3.11 Model with a Binary Factor (Two Subpopulations): Two-Sample Independent t

A subset of information from the **Hospital Doctor Patient (HDP)** data is used to demonstrate the fit of the models (see https://stats.idre.ucla.edu/r/codefragments/mesimulation/#tumor). These data are obtained on a three-level hierarchical structure with patients nested within doctors, and doctors nested within hospitals. The data in HDP include two outcomes: tumor size in millimeters (mm) and CO_2 levels in percentage. Tumor size and CO_2 are indicators of disease severity and are highly correlated. The analysis of these data ignores the hierarchical structure of the data. Therefore, these responses are treated as though they come from an independent process. A subset of the data is provided in Table 3.4.

The distribution of tumor sizes for females and the distribution for males are plotted in Figure 3.3. The data seem to be distributed differently between males and females. Each graphs looks like they are symmetric mound-shaped curves. It appears in the graph for females, that females tend to have a larger concentration near the mean (kurtosis). There is also more "peakedness" among the females. We explore the difference between females and males.

3.12 Modeling Continuous Response with a Binary Factor

In this section, statistical models for continuous responses and binary responses, with a single binary factor (two subpopulations) in the model,

TABLE 3.4

Subset of HDP Data

Tumor Size	Co2	Pain	Married	Smoking	Gender	Cancer Stage	BMI
67.98	1.53	4.00	0.00	former	male	II	24.14
64.70	1.68	2.00	0.00	former	female	II	29.41
51.57	1.53	6.00	1.00	never	female	II	29.48
86.44	1.45	3.00	0.00	former	male	I	21.56
53.40	1.57	3.00	0.00	never	male	II	29.82
51.66	1.42	4.00	1.00	never	male	I	27.10
78.92	1.71	3.00	1.00	current	female	II	21.12
69.83	1.53	3.00	0.00	former	male	II	42.48
62.85	1.54	4.00	1.00	former	male	II	18.69
71.78	1.59	5.00	0.00	never	male	II	39.44

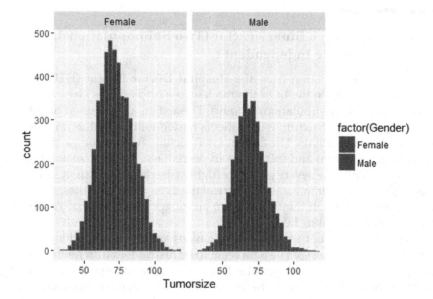

FIGURE 3.3
Graph of Tumor sizes for Females and Males

are presented. As such, the methods concentrate on the mean parameter of a continuous response across two subpopulations and the mean for binary responses also across two subpopulations. These models compare the means (called proportions if binary outcomes) across two groups. In our example, in particular, interest is on tumor sizes to see if there is a differential effect based of gender. Both parametric (assume the distribution of the response is known) and nonparametric (when the distributional assumptions are not known) methods are presented.

3.12.1 Parametric Tests for Population Mean Difference

Suppose there is an interest in knowing if tumor sizes differ between males and females. We model the mean parameter of the distribution for each group. Assuming the population of tumor sizes follows a normal distribution for each group, we consider a parametric approach. Thus, we test the hypotheses:

$$H_0 : \mu_{males} = \mu_{females}$$

$$H_1 : \mu_{males} \neq \mu_{females}.$$

where μ_{males} and $\mu_{females}$ are the true mean of tumor sizes among males and females.

For these hypotheses, consider a statistical test to compare the population means from the two distributions. One can present this research question in some related forms. One may be interested in knowing if gender has an impact on tumor size, or if there is a differential effect due to gender on tumor size. Thus, in each case, our research hypothesis, H_1, will target the means of the two distributions – one for males and one for females. Our binary factor is gender. Though we are addressing individuals, rather, we are modeling the means of tumor size from these individuals.

3.12.2 Assumptions

There are three basic assumptions one must satisfy to use this test appropriately:

1. Assume that each population distribution is normal (or Gaussian). If a test shows that the observations are not normal, then one can use a transformation of the tumor sizes (but no longer a common approach), or one can fit a generalized linear model (common approach), or one can use a nonparametric test;

2. Assume that the variability among tumor sizes in males is the same as the tumor sizes among females. This assumption is referred to as homoscedasticity (common variance). We can test for homoscedasticity, and if the evidence suggests the variance is not constant, then one can use the Satterthwaite's approximation (Satterthwaite 1946) or the Levine test (Brown and Forsythe 1974) approximation;

3. Also, we assume that the observations are independent. By independence, it means that the mechanism that gives rise to each outcome is "memoryless" when giving rise to other outcomes.

3.12.3 Two-sample Independent t-Statistic

To address the research question, one needs to construct a measure that takes the key information (mean, sample size, standard error) from the two samples of respondents. One can construct an overall two-sample test of the difference in proportion values. Based on the size of that test statistic value, one can make a decision on the research hypothesis and give the researcher a sense of belief or doubt about the hypothesis. We compute the two-sample independent t-statistic. We refer to it as independent, as there are two samples from two different populations. A measure of that difference is reflected in the value of the statistic. There is no reason to believe that there is a commonality among outcomes in reference to tumor size. It is useful to note that the two-sample independent t-test statistic behaves as though there are no other factors (other than the binary factor under consideration) to influence the response. This is an example of two-at-a-time models with one factor. The interpretation must be aligned in terms of the model. This is different from a multivariable model. We encourage readers to visit Arreola Vazquez, Irimata, and Wilson (2020).

Consider the model:

$$\text{Observation}_{ij} = Y_{ij} = \mu_j + \text{Error}_{ij}$$

where Observation_{ij} denotes the value from the ith subject in the jth group and μ_j is the mean of the jth group, with Error_{ij} representing the difference between the observation and its group mean. In this, case $j = 1, 2$[male and female]. The model can be equally represented as

$$\text{Observation}_{ij} = Y_{ij} = \mu + \alpha_j + \text{Error}_{ij}$$

where Observation_{ij} denotes the value from the ith subject in the jth group, α_j is the jth group extra effect above or below the mean due to the jth group, and Error_{ij} represents the difference between the observation and its group mean. Both parameterizations are equivalent. The first parameterization accommodates tests between the two group means $\mu_1 = \mu_2$ while the second parameterization represents tests between the group effects, so

$$\mu_1 = \mu + \alpha_1,$$

$$\mu_2 = \mu + \alpha_2$$

which implies that $\alpha_1 = \alpha_2$ and μ is the common mean for both groups. Thus testing the equal means is equivalent to testing the equal group effects.

3.12.4 Decision Using Two-Sample t-Test

The size of the statistical test value t (that measures the difference in means between the two groups) is referred to as the independent two-sample t test:

$$t = \frac{\text{Diff}_{(\text{sample mean}_{grp1} - \text{sample mean}_{grp2})}}{\text{Standard deviation}_{\text{Diff}_{(\text{sample mean}_{grp1} - \text{sample mean}_{grp2})}}}.$$

One needs a method to tell whether the difference between group means given by the t-statistic is large enough to conclude that there is sufficient evidence to support the research statement. We rely on the distribution that the t-test statistic follows. In this case, it is a t-distribution with degrees of freedom equal to the total sample size minus two. (Degrees of freedom is a parameter of the t distribution. It represents the sample size minus having to estimate two parameters-population means.) A p-value is obtained, based on the assumed t distribution related to the test statistic. The p-value (strength of the data against the null hypothesis) gives a sense that one can support or refute the null hypothesis. There is a need for a cut point to determine when the p-value is small enough to make a decision in favor of the research hypothesis. The cut point is referred to as the critical value corresponding to the significance level or risk level or α level (probability of rejecting the null hypothesis when the hypothesis is true). The critical value is found from the test statistic distribution as the point on the distribution, equivalent to the area under the curve in the tail. It is a value set by the researcher in conjunction with the client before conducting the experiment.

3.13 Analysis of Data for Two-Sample t-Test in SAS and R

The Hospital Doctor Patient data are used for testing that the mean tumor size differs based on gender. We present results obtained with two statistical software, SAS and R.

3.13.1 Analysis of Data with SAS Program

```
proc glm data =tumorsize;
class Gender;
model tumorsize=Gender;
run;
proc ttest data=tumorsize;
class Gender;
var tumorsize;
```

66Statistical Analytics for Health Data Science with SAS and R

```
run;
proc sgplot data=tumorsize;
scatter x=Gender y=tumorsize;
run;
The SAS code presents the output.
```

The Two Sample Independent t-TEST Procedure
Variable: tumorsize

Sex	N	Mean	Std Dev	Std Err	Minimum	Maximum
Female	5115	72.083	12.338	0.173	35.187	116.5
Male	3410	69.077	11.420	0.196	33.969	112.6
Diff (1–2)		3.0055	11.9789	0.2648		

Sex	Method	Mean	95% CL Mean		Std Dev	95% CL Std Dev	
Female		72.083	71.745	72.421	12.3375	12.103	12.581
Male		69.077	68.694	69.461	11.4197	11.155	11.698
Diff (1–2)	Pooled	3.006	2.486	3.525	11.9789	11.802	12.161
Diff (1–2)	Satterthwaite	3.006	2.494	3.517			

Method	Variances	DF	t Value	Pr > |t|
Pooled	Equal	8523	11.35	<.0001
Satterthwaite	Unequal	7679	11.53	<.0001

The output gives summary statistics for each group. It provides the differences for both groups. Two sets of results are given. One set if the variances of the subpopulations are equal (Pooled) and another set if the variances of the subpopulations are unequal (Satterthwaite). In both cases the results show significant differences between the means t = 11.35 p <.0001 for Pooled and t = 11.53 p <.0001 for Satterthwaite.

3.13.2 Analysis of Data with R Program

```
## Call R function "lm"
lm = lm(tumorsize~factor(Gender), data=tumorsize)
summary(lm)
##
## Call:
## lm(formula=tumorsize~factor(Gender), data= tumorsize)
##
## Residuals:
##        Min       1Q    Median       3Q       Max
## -36.896    -8.359    -0.788    7.960    44.375
##
## Coefficients:
```

```
##                   Estimate Std. Error t value    Pr(>|t|)
## (Intercept)        72.0829     0.1675  430.37     <2e-16 ***
## factor(Gender)Male -3.0055     0.2648  -11.35     <2e-16 ***
## Signif. codes:0 '***' 0.001 '**' 0.01 '*' 0.05 '.' 0.1 ' ' 1
```

$\hat{\mu}_{tumor\ size} = 72.0829 - 3.0055 Males$ When Males = 1, then the mean is 69.077; When males = 0, then we have mean for females = 72.0829. The 72.0829 is the mean for the female, and the males are −3.0055 less than the females. The p-value for the factor (gender) Male is <2e-16. This tells that the mean for males is significantly different from females.

```
## Residual standard error: 11.98 on 8523 degrees of
               freedom
## Multiple R-squared: 0.01489,   Adjusted
               R-squared: 0.01477
## F-statistic: 128.8 on 1 and 8523 DF, p-value: <
               2.2e-16
```

We look at the test for difference between males and females. We see that p-value: < 2.2e-16 with F = 128.8, thus the mean tumor size differs between males and females. Thus, we conclude there are significant differences in tumor sizes based on gender.

```
Aov = aov(tumorsize~factor(Gender), data = tumorsize)
summary(aov)
```

```
##                 Df    Sum Sq    Mean Sq F value Pr(>F)
## factor(Gender)  1     18482     18482   128.8   <2e-16 ***
## Residuals       8523  1222993   143
## ---
## Signif. codes:    0 '***' 0.001 '**' 0.01 '*' 0.05 '.' 0.1 ' ' 1
```

t.test(tumorsize~Gender,tumorsize,var.equal=TRUE) # student t-test
   ##

We obtain the same results with aov and the t-test commands. We obtain the test for difference between males and females. We see that p-value: < 2.2e-16 with F = 128.8, thus the mean tumor size differs between males and females. Thus, we conclude there are significant differences in tumor sizes based on gender.

```
   ## Two Sample t-test
##
## data: tumorsize by Gender
## t = 11.349, df = 8523, p-value < 2.2e-16
## alternative hypothesis: true difference in means is
not equal to 0
## 95 percent confidence interval:
## 2.486378 3.524630
## sample estimates:
## mean in group Female   mean in group Male
##           72.08287        69.07737


t.test(tumorsize~Gender,tumorsize)   # Welch t-test
## Welch Two Sample t-test
## data: tumorsize by Gender
## t = 11.525, df = 7679, p-value < 2.2e-16
## alternative hypothesis: true difference in means is
not equal to 0
## 95 percent confidence interval:
## 2.494320 3.516687
## sample estimates:
## mean in group Female   mean in group Male
##      72.08287       69.07737
```

> We look at the test for difference between males and females. We see that p-value: < 2.2e-16, thus the mean tumor size differs between males and females. Thus, we conclude there are significant differences in tumor sizes based on gender.

Comment: We conducted a statistical test to compare mean tumor sizes between males and females. We use two (2) different statistical software packages, SAS and R, to analyze these data. We obtain a p-value of <.0001. Thus, we conclude that there is sufficient evidence to say that the mean tumor sizes differ between males and females. If the interest is in the mean tumor sizes for females being greater than males (consider directional one-tailed test) then (if p-values were provided on the statistical program by default) one would take the p-value of 0.0001 divided by 2. therefore, we conclude that the mean tumor sizes for females are greater than the mean tumor sizes for males. It is important to emphasize that the two-sample independent t-test is a univariate test (only one response, tumor size (continuous variable) with one binary factor (gender)). In addition, it is not multivariable (as it has only one factor, gender). The hypothesis accounts for only one factor, gender, to explain differences in tumor size. It is rarely the case in research, especially with observational studies, that one factor is sufficient to explain all the variation in the responses. In later chapters, there are cases for addressing multiple factors through the use of covariates. In such cases, the simultaneous effect of the covariates can lead to different conclusions.

3.14 Nonparametric Two-Sample Test

When we use the two-sample independent t-test, we assume that the population of outcomes (tumor sizes) in each group is normally distributed. It is not uncommon at times, however, to find that the distributional assumption is not met. For example, when income is the outcome variable, it may be possible to resort to a nonparametric test, as the normal assumption is not necessarily satisfied.

3.14.1 Analysis of Data for Two-Sample Nonparametric Test

We demonstrate the use of several nonparametric tests, the Van der Waerden two-sample test in SAS and n R. This nonparametric test relies on two assumptions: the data are ranked, and the observations are independent. The ranked data are used to obtain a statistical test similar to our construction of a parametric test with a reference distribution culminating in a p-value. We present results using the nonparametric tests based on its ranking of tumor size for each group (males and females). The outputs from two statistical software programs, SAS and R are presented in full, or in some cases, a truncated version.

3.14.2 Analysis of Data SAS Program

```
proc NPAR1WAY data=tumorsize normal;
class Gender;
var tumorsize; run;
The SAS code provides the following output
```

NPAR1WAY Procedure

			Van der Waerden Scores (Normal) for Variable tumorsize		
			Classified by Variable Gender		
Gender	N	Sum of Scores	Expected Under H0	Std Dev Under H0	Mean Score
Male	3410	−499.73193	0.0	45.192902	−0.146549
Female	5115	499.73193	0.0	45.192902	0.097699

Van der Waerden Two-Sample Test	
Statistic	−499.7319
Z	−11.0578
One-Sided Pr < Z	<.0001
Two-Sided Pr > |Z|	<.0001
Pr > Chi-Square	<.0001

Testing difference between males and females under the non-normality assumption: Van der Waerden two-sample test has p-value: <0.0001, thus the mean tumor size differs between males and females. Then it is concluded there are significant differences in tumor sizes based on gender.

Van der Waerden One-Way Analysis	
Chi-Square	122.2739
DF	1
Pr > Chi-Square	<.0001

A test for difference using the Van der Waerden one-way analysis with p-value: <0.0001. We provide two tests, the Z test [−11.0578] and the chi-square test [122.739]. Take Z^2, we get the chi-square.

3.14.3 Analysis of Data R Program

```
wilcox.test(tumorsize$tumorsize) #Wilcoxon sign rank test
##
## Wilcoxon signed rank test with continuity correction
##
```

```
## data: tumorsize$tumorsize
## V = 36342000, p-value < 2.2e-16
## alternative hypothesis: true location is not equal to 0
```

> We test for difference between males and females under the non-normality assumption of the distribution. The Wilcoxon signed-rank test has p-value: $<2.2\times10^{-16}$; thus, the mean tumor size differs between males and females. Thus, we conclude there are significant differences in tumor sizes based on gender.

3.15 Modeling on Categorical Response with Binary Covariate

In this section, we consider a categorical response in cancer stage with a binary factor in marital status (single versus married). We are not implying a causal effect, but rather using the binary factor as a measure of comparison of cancer stage to demonstrate the use of a binary factor on a categorical response. As the response is categorical, these data are presented into a 2 by 4 contingency table (rectangular array of cell counts), as seen in Table 3.5.

In Table 3.5, the two rows account for marital status (0 = no, 1 = yes) and four columns to identify the cancer stages. The cells represent counts or the frequency of cases with the set of characteristics of marital status by cancer stage. The response (cancer stage) has more than two categories, so we consider the multinomial distribution (if we had two possible outcomes, we would call it binomial) to analyze these data. It is important to know how the data are collected, as such information leads to different hypotheses. For example, were these data obtained based on known group sizes? Then, we have a test of homogeneity, a comparison of the corresponding probabilities over the groups. To identify the differences in these proportions, one will need to construct a distance measure that takes the key information about sample information (cell proportion, row totals, column totals, and

TABLE 3.5

Cross Classification of Marital Status by Cancer Stage

Marital Status	I	II	III	IV	Row Total
0	836	1378	753	443	3410
1	1722	2031	952	410	5115
Column Total	2558	3409	1705	853	8525

sample size) between married and unmarried from the 8525 patients into a single value. An overall index that measures the distance between observed cell values and what we expect is needed if the null hypothesis is true. The desired index is the statistical test value. Based on the size of the test statistic value, one makes a decision on the null hypothesis. This distance measure is the Pearson chi-square test statistic. There are other measures, such as, G^2 and Neyman statistic, to name a couple. The distance pertains to the difference between observed value and the expected value (under H_0). Reference to some tests in our analyses of the data using different statistical software programs is made in SAS and R.

Suppose the data were obtained with no knowledge of marital status, except the knowledge of the overall sample size of 8525 individuals. Then we have a test of independence.

$$H_0 : P_{ij} = P_{i+}P_{+j}$$

$$H_1 : P_{ij} \neq P_{i+}P_{+j}$$

In this case, for both hypotheses, test of independence and test of homogeneity, will result in the same p-values. However, the interpretation differs. In both cases, a distance measure (statistical test value) needs a method to tell when the measure is large enough to declare that there is sufficient evidence to support the research question. We use the distribution that the distance measure follows to obtain a p-value. In this case, the distribution is chi-square with certain degrees of freedom as depicted in Figure 3.4. While we obtain the p-value based on the distribution of the distance measures, there is a need for a cut point to determine if it is small enough to make a decision to declare that the research statement is supported. This cut point is referred to as the critical value and is provided by the significance level or α level or risk level. One should report the p-value rather than compare it with the risk level, α. The p-value is a conditional probability. The smaller its

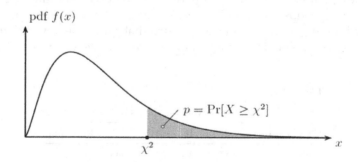

FIGURE 3.4
Illustration of Chi-Square Distribution with large degrees of freedom

value, the less we are inclined to believe our null hypothesis. At such time, one takes the position of the research hypothesis and declares that there are significant results. In recent times, the American Statistical Association has made a statement about the reporting of p-values, (see Wasserstein and Lazar 2016).

> **Comment:** In this example, an asymptotic test is used, as we had a very large number of patients (8525). The response variable has four stages of cancer (I, II, III, and IV). The Pearson chi-square test is non-directional, and it is not the best way to analyze these data if one has questions about certain categories. One can use the multinomial logistic regression model, which allows access to more involved information, or one can fit a log-linear model. Such models are presented in Chapter 8.

3.15.1 Analysis of Data

We consider two procedures to test the hypothesis regarding marriage and cancer stage. We refer to these as the test of proportions and the test of homogeneity, respectively.

$$H_{0:pro} : P_{stage\,I} = P_{stage\,II} = P_{stage\,III} = P_{stage\,IV}$$

$$H_{0:homog} : \begin{pmatrix} P_{I|NM} \\ P_{II|NM} \\ P_{III|NM} \\ P_{IV|NM} \end{pmatrix} = \begin{pmatrix} P_{I|M} \\ P_{II|M} \\ P_{III|M} \\ P_{IV|M} \end{pmatrix}$$

The analysis of contingency tables is in keeping with the measures of association, which are usually done through log-linear models, where the log of the cell count is modeled without explanatory factors (Bishop, Feinberg, and Holland 1975).

3.15.2 Analysis of Data with SAS Program

*** Statistical Tests for Categorical data;

```
proc catmod data=tumorsize;
model cancerstage=Married;
run;
```

```
proc freq data=tumorsize;
table cancerstage*Married/all;
run;
proc catmod data=tumorsize;
model cancerstage=;
run;
```

The CATMOD Procedure

Population Profiles		
Sample	Married	Sample Size
1	0	3410
2	1	5115

The test in which proportions are the same across the four stages (Ho) is tested by the chi-square test on intercept line with p <.0001.

The test of homogeneity in which proportions are the same across the four stages between married and not married is tested by the chi-square of 118.74 with 3 df on "married" line with p <.0001.

Maximum Likelihood Analysis			
Source	DF	Chi-Square	Pr > ChiSq
Intercept	3	1462.20	<.0001
Married	3	118.74	<.0001
Likelihood Ratio	0	.	.

Analysis of Maximum Likelihood Estimates						
Parameter		Function Number	Estimate	Standard Error	Chi-Square	Pr > ChiSq
Intercept		1	1.0351	0.0402	662.03	<.0001
		2	1.3675	0.0385	1264.69	<.0001
		3	0.6865	0.0421	266.42	<.0001
Married	0	1	-0.4000	0.0402	98.87	<.0001
	0	2	-0.2327	0.0385	36.61	<.0001
	0	3	-0.1560	0.0421	13.75	0.0002

The parameter (intercept) tells that $P_{stage\ I}$ is different from $P_{stage\ IV}$ and so is $P_{stage\ II}$ and $P_{stage\ III}$. The three p-values are <.0001.

The parameter (married) tells that $P_{stage\ I|married}$ differs from $P_{stage\ I|\ not\ married}$. This is also the case for stage II and stage III.

The data for a binary variable influencing a categorical response with four categories present a contingency 2×4 table. We always want the observed cell values to be large enough because the statistics rely on asymptotic properties.

Statistics for Table of Cancer Stage by Married			
Statistic	DF	Value	Prob
Chi-Square	3	120.2762	<.0001
Likelihood Ratio Chi-Square	3	120.4353	<.0001
Mantel-Haenszel Chi-Square	1	117.0642	<.0001
Phi Coefficient		0.1188	
Contingency Coefficient		0.1180	
Cramer's V		0.1188	

The test for homogeneity is the same in value as the test of independence. There are three such tests given in SAS. They gave p-values of <.0001, thus suggesting that homogeneity does not hold.

Statistic	Value	ASE
Gamma	−0.1838	0.0168
Kendall's Tau-b	−0.1073	0.0099

SAS provides a series of measures of association. They do not provide p-values, but one can obtain a Z-value through the ratio of value to ASE. These measures are suited based on the ordinal nature of one or both variables.

3.15.3 Analysis of Data with R Program

```
## Load the "nnet" library
library(nnet)
## Call "multinorm" function
multi1 = multinom(CancerStage~Married,data=tumorsize)
```

```
## # weights: 12 (6 variable)
## initial value 11818.159429
## iter 10 value 10851.379037
## final value 10851.376290
## converged

summary(multi1)

## Call:
## multinom(formula = CancerStage ~ Married, data =
##                     tumorsize)
##
## Coefficients:
##          (Intercept)     Married
## II        0.4997634    -0.3346969
## III      -0.1045543    -0.4880974
## IV       -0.6350558    -0.7999737
##
## Std. Errors:
##          (Intercept)     Married
## II        0.04383910    0.05472625
## III       0.05024133    0.06446189
## IV        0.05876661    0.08045609
##
## Residual Deviance: 21702.75
## AIC: 21714.75

library(stargazer)
## Warning: package "stargazer" was built under R
##                     version 3.3.2
##
## Please cite as:
## Hlavac, Marek (2015). stargazer: Well-Formatted
##           Regression and Summary Statistics Tables.
## R package version 5.2. http://CRAN.R-project.org/
##           package=stargazer
stargazer(multi1, type="text")

##
## ======================================================
##              Dependent variable:
##          ---------------------------------
##              II     III    IV
##             (1)    (2)    (3)
##          ------------------------------------------------------
```

```
## Married    -0.335***   -0.488***   -0.800***
##            (0.055)     (0.064)     (0.080)
##
## Constant   0.500***    -0.105**    -0.635***
##            (0.044)     (0.050)     (0.059)
##
## -----------------------------------------------------
## Akaike Inf. Crit. 21,714.750 21,714.750 21,714.750
## =====================================================
## Note:       *p<0.1; **p<0.05; ***p<0.01
```

> The test for homogeneity is equal to the test of independence. They
> gave p-values virtually equal to 0, thus suggesting that homogeneity
> does not hold, or there is a relationship between marital status and
> cancer stage.

3.16 Summary and Discussion

This chapter concentrates on statistical models with a binary factor. Both the
two-sample independent t-test for comparing means and the test of homogeneity or proportions (for testing binary responses) indicate that the binary
factor is significant. However, these models only considered one factor. In
practice, a system consists of several covariates that simultaneously may
influence the response. It is important that researchers understand that these
one-factor models are good to help build to a multivariable model. They are
not best in isolation except in experimental conditions. The key fact is that
the role of the factor may change when analyzed alone as opposed to in a
system of factors. More so, the interpretation differs between a multivariable
and a two-at-a-time data analysis.

   Question: If we had only 40 observations and had to address the
   hypothesis that marital status was related to cancer stage, how
   would one analyze these data?

   Answer: There are $2 \times 4 = 8$ cells. Since there are 40 observations, the
   best situation one can have for cell sizes would be $40/8 = 5$. Five
   is about the smallest value one wants in these cell counts if one is
   going to use the Pearson or likelihood ratio chi-square test. With
   such small sample cell sizes, one should use the multinomial version
   of Fisher's exact test.

3.17 Exercises

Using the German Breast Cancer data, answer the following questions.

1. Obtain a plot of the size of nodes similar to Figure 3.1.
2. Is the size of nodes distributed normally?
3. Is the mean of the size of the nodes greater than 25?
4. Is the percentage of grade 1 larger than 0.75?

Using the HDP dataset, answer the following:

1. Is there a differential effect based on gender for BMI?
2. Is there homogeneity across gender?
3. How does gender relate to cancer stages?
4. Use Fisher's exact test to check the association between gender and smoking.

References

Armitage, P., Berry. G.: Statistical Methods in Medical Research, 4th ed. Blackwell Scientific Publications, Oxford (2008)

Arreola, E. V., Irimata, K., Wilson, J. R.: Common errors of interpretation in biostatistics. Biostatistics and Epidemiology, 238–46 (2020). https://doi.org/10.1080/24709360.2020.1790085

Bishop, Y.M., Feinberg, S.E., Holland, P.W.: Discrete Multivariate Analysis: Theory and Practice. Springer, New York (1975)

Brown, M.B.: Forsythe, Robust tests for the equality of variances. A. B. Journal of the American Statistical Association, 69, 364–7 (1974)

Clopper, C.J., Pearson, E.S.: The use of fiducial limits illustrated in the case of the binomial. Biometrika, 26, 404–13 (1934)

Greenland, S., Senn, S.J., Rothman, K.J., Carlin, J.B., Poole, C., Goodman, S.N., Altman, D.G.: Statistical tests, P values, confidence intervals, and power: a guide to misinterpretations. European Journal of Epidemiology, 31(4), 337–50 (2016)

Michaud, L.B.: Treatment-experienced breast cancer. American Journal of Health-System Pharmacy, 65(10 Suppl 3), S4–9 (2008, May 15). https://doi.org/10.2146/ajhp080088. PMID: 18463331.

Randles, R.H., Wolfe, D.A.: Introduction to the Theory of Nonparametric Statistics. John Wiley & Sons, Hoboken (1979)

Sauerbrei, W., Royston, P.: Building multivariable prognostic and diagnostic models: Transformation of the predictors by fractional polynomials. Journal of the Royal Statistical Society, Series A, 162, 71–94 (1999). https://doi.org/10.1111/1467 -985X.00122

Satterthwaite, F.E.: An approximate distribution of estimates of variance components. Biometrics Bulletin 2(6), 110–4 (1946)

Schmoor, C., Olschewski, M., Schumacher, M.: Randomized and non-randomized patients in clinical trials: Experiences with comprehensive cohort studies. *Statistics in Medicine*, 15(3), 263-71 (1996). https://doi.org/10.1002/(SICI)1097-0258(19960215)15:3<263::AID-SIM165>3.0.CO;2-K. PMID: 8643884.

Snedecor, G., Cochran, W.: Analysis of Variance: The Random Effects Model. Statistical Methods. Iowa State University Press, Ames, IA, 237–53 (1989)

Wasserstein, R. L., Lazar, N. A.: The ASA statement on p-values: context, process, and purpose. The American Statistician, 70(2), 129–33 (2016)

4

Modeling of Continuous and Binary Outcomes with Factors: One-Way and Two-Way ANOVA Models

4.1 Research Interest/Question

In Chapter 3, we noted that researchers are often interested in determining whether one method is better than another method or a treatment is better than a control. To make their determination, they usually rely on the means or medians between the two groups. However, there are situations when one may be interested in comparisons for more than two groups. In this chapter, we concentrate on statistical models for comparing more than two groups for the case of continuous outcomes and for the case of binary outcomes. In other words, the focus in this chapter is on cases when the variable of interest is impacted by one categorical factor with more than two levels.

In addition, the simultaneous impact of two categorical factors is also considered. When there is one categorical factor and one continuous outcome, we examine with a one-way analysis of variance (ANOVA) model. When there are two simultaneous categorical factors for one continuous outcome, we examine with a two-way ANOVA model.

We demonstrate these ANOVA models with the **Hospital, Doctor, and Patient** (HDP) dataset used in Chapter 3. We focus on tumor size as the outcome variable. We examine a one-way ANOVA model with smoking as the factor, and two-way ANOVA model with smoking and cancer stage as simultaneous factors.

4.2 Hospital Doctor Patient: Tumor Size

In the HPD dataset, the observations are simulated as a three-level hierarchical structure with patients nested within doctors and doctors nested within

DOI: 10.1201/9781003315674-4

TABLE 4.1

Subset of HDP Dataset

tumorsize	co2	remission	Age	SmokingHx	Sex	CancerStage	BMI
67.98	1.53	0.00	64.97	former	male	II	24.14
64.70	1.68	0.00	53.92	former	female	II	29.41
51.57	1.53	0.00	53.35	never	female	II	29.48
86.44	1.45	0.00	41.37	former	male	I	21.56
53.40	1.57	0.00	46.80	never	male	II	29.82
51.66	1.42	0.00	51.93	never	male	I	27.10
78.92	1.71	0.00	53.83	current	female	II	21.12
69.83	1.53	0.00	46.56	former	male	II	42.48
62.85	1.54	0.00	54.39	former	male	II	18.69
71.78	1.59	0.00	50.54	never	male	II	39.44

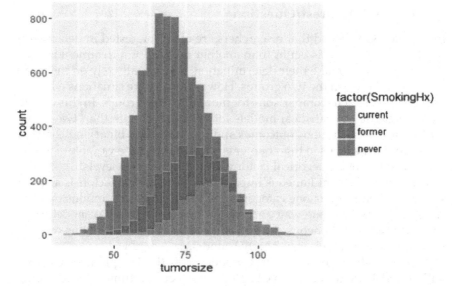

FIGURE 4.1

Graph tumor size by smoking history

hospitals. These data are obtained for the outcomes of tumor size in millimeters (mm) and CO_2 levels in percentage. Tumor size and CO_2 are both measures of disease severity and are highly correlated. In this chapter, we analyze these data while ignoring the structure under which the data are obtained. In fact, these observations are treated as though they are independent. A sample of these data is given in Table 4.1.

A distribution of tumor size by smoking history is given in Figure 4.1. The smoking history is categorized as former smoker, never smoked, and current smoker. The mean tumor sizes for these subgroups are displayed in Figure 4.2.

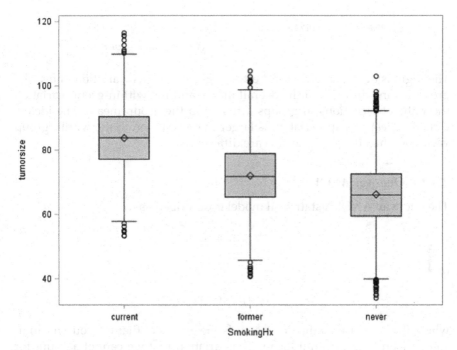

FIGURE 4.2
Graphical display by smoking history

4.3 Statistical Modeling of Continuous Response with One Categorical Factor

4.3.1 Models

In this section, we explain how to compare group means for more than two groups. To test whether the means of continuous outcomes are statistically different across groups, one fits a one-way analysis of variance (ANOVA) model. It is limited in that it will only tell if the means are equal or not. A one-way ANOVA model compares the means through the partitioning of the variance of the outcomes. The partition consists of splitting the overall variation into "between-group" and "within-groups". Thus, for each observation:

$$[observation - overall\,mean] = [observation - group\,mean] \\ + [group\,mean - overall\,mean]$$

Squaring both sides and summing (the cross product can be proven to be zero), resulting in

$$\sum [observation - overall\, mean]^2 = \sum [group\, mean - overall\, mean]^2$$
$$+ \sum [observation - group\, mean]^2$$

The right side represents the between-group means variation (how the group means differ from the overall mean) and the within-group variation (how the observations in groups differ from the group mean). The idea is if the between-group variation is larger in comparison to the within-group variation, then the group means are different.

4.3.1.1 One-way ANOVA

The one-way ANOVA statistical model is described as

$$Y_{ij} = \mu_i + \varepsilon_{ij}$$

or

$$Y_{ij} = \mu + \alpha_i + \varepsilon_{ij}$$

where the i represents the groups, α_i is the effects of that i^{th} group mean in comparison to the overall mean. The variation that we cannot account for within the group means is noted as unaccounted for through the random term, ε_{ij}. We assume that there is no systematic variation within ε_{ij} but if there is systematic variation, then one wants to identify the covariate that produces the pattern and include it in the systematic portion. However, we assume that this random portion unaccounted for follows a standard normal distribution, then the outcome j for group i is

$$Outcome_{ij} = overall\, mean + group_{effect\, i} + random_{ij}$$

The statistical model

$$y_{ij} = \mu_i + \varepsilon_{ij}$$

where y_{ij} denotes the jth unit within the ith group, μ_i is the mean of the ith group, μ is the overall mean across all groups and ε_{ij} is a measure of the unexplained variation for the jth unit in the ith group, as noted previously. We assume that random variable ε_{ij} is distributed normally with mean zero and variance σ^2, which means that observations y_{ij} come from a normally distributed population with mean μ_i and variance σ^2. A reparametrization (presenting in an alternative way) of $\mu_i = \mu + \alpha_i$ means that instead of interest in the group means, we are interested in the deviations α_i from the group mean. These $\pm_i i = 1, 2, \ldots I$; are referred to as group effects.

4.3.1.2 F-Statistic in One-Way ANOVA

To answer the research question, one needs to construct a measure that takes the key information (group means, overall group means, sample size, standard error) from the sample of patients into account. It is a test statistic. Based on the size of the test statistic, one can make a decision about the hypothesis or give the researcher a sense about the beliefs or disbeliefs of the research hypothesis. In particular, the test statistic is referred to as the one-way ANOVA F-statistic. One refers to this model as a one-way ANOVA, as there is only one categorical variable as a factor (i.e., smoking history). This categorical variable allows us to classify the data into between-group variation and within-group variation. The test statistic that compiles all the information from the sample for this one-way ANOVA model is the F-test statistic. If, for example, one had considered gender as well as smoking history, then one would consider a two-way ANOVA model.

The ANOVA F-test tells if there is a significant difference among group means but does not say which one of the group means is significantly different from another. In fact, the model is testing the null hypothesis $H_0 : \mu_{cur} = \mu_{for} = \mu_{nev}$ versus H_1 not all the means are equal, where μ_{cur} is the true mean tumor size among current smokers, μ_{for} is the true mean tumor size among former smokers and μ_{nev} mean tumor size among those who never smoked. This is important if one wants to know if a group mean tumor size differs as compared to a particular smoking group (former smoker, current smoker, or never smoked). Alternatively stated, one may be interested in knowing if smoking has an impact on tumor size or if there is a differential effect due to smoking history on tumor size. Thus, our research hypothesis, H_1, will target the means of the distributions for the three subpopulations – one for former smokers, one for current smokers, and one for those who never smoked. The factor of interest is smoking history. The data for tumor sizes are summarized in Table 4.2. These summary statistics are not sufficient to tell if the population means are equal. To answer these questions, one uses a one-way ANOVA F-test. Table 4.2 shows three levels in the group

TABLE 4.2

Descriptive Statistics by Groups

Group	N	Mean	Std. Deviation	Std. Error	95% Confidence Interval for Mean	
					Lower Bound	Upper Bound
Current	1705	83.773	9.993	.242	83.299	84.248
Former	1705	72.022	10.034	.243	71.545	72.499
Never	5115	66.203	9.910	.139	65.931	66.474
Total	8525	70.881	12.068	.131	70.624	71.137

variable smoking. There are 8525 observations. There are 1705 patients who currently smoked and 1705 who are former smokers.

4.3.1.3 Assumptions

Assume for each subpopulation (current, former, and never smoker) tumor sizes are distributed as normal (or Gaussian). The assumption is about the population, although there is only a sample of observations available from each subgroup. If the test shows that they are not normal, then one can use a transformation of the tumor sizes (but no longer a common approach), or one can fit a generalized linear model (*common approach*), or one can use a non-parametric test (presented later in this section). Also assume that the variability among the tumor size populations in former, current, and never smoked are the same (homoscedasticity). One can test for equality of variances, and if they are not (variance not constant is referred to as heteroscedasticity), one uses a weighted least squares or some other model that addresses heteroscedasticity. Further, assume that the observations are independent.

4.3.1.4 Decision

Once the size of the *F test statistic* value is obtained, one needs a method to determine when the value is large enough to decide that there is sufficient evidence to support the statement in the research hypothesis. Use the distribution that the test statistic is known to follow. In this case, it follows an F distribution, (F-distribution depends on the numerator and denominator degrees of freedom). Obtain a p-value, which depends on the assumed F distribution of the test statistic values. The p-value gives us a sense of belief as to which hypothesis (H_0 or H_1) is true. The p-value needs a cut point to decide if it is small enough to support a decision for H_1. The cut point is denoted as the point where the area under the tail of the F-distribution is the significance level or α.

4.3.1.5 Multiple Comparisons

While the one-way ANOVA model tells us whether the group means are equal or different, one may have further questions about relationships among the individual group means. For example, is the mean of subgroup I the same as subgroup II but different from that of subgroup III? Such questions arise when observing more than two groups, and there is a natural inclination to examine them *two-at-a-time*. Such an approach of two-at-a-time is not supported in the statistical literature without paying attention to the multiple comparisons of group means.

Each of these two-at-a-time comparisons results in an increase in the probability of committing a type I error. Since each comparison inflates the probability of a type I error (rejecting H_0 and concluding H_1 is true, when in fact H_0 is true), there is a need to adjust for the inflation of the error. This leads to the use of post-hoc tests to control the magnitude of the type I error (often called the experiment wise or family wise error) and make multiple comparisons. The major issue in any discussion of multiple-comparison procedures is the question of the probability of type I error.

Day and Quinn (1989), among others, pointed out the purpose of the multiple-comparison procedures is to control the overall significance level. They explained that the overall significance level or error rate is the probability, conditional on all the null hypotheses being tested is true, of rejecting at least one of them. Many multiple comparison procedures have been proposed for controlling the overall type I error. Some are for designed comparisons among means: orthogonal contrasts (with F tests), Bonferonni adjustment, and Scheffe's interval; and others are post-data comparisons: Scheffe's interval, all pairwise comparisons, Student-Neuman-Keuls (SNK) test, and Tukey's honest significant difference (HSD); while yet others are comparisons with control: Dunnett's test and William's test. Whereas all control the overall Type I error, it should be pointed out that each contrast has a type I error level of α (pay no penalty in terms of type I error) for **orthogonal multiple comparisons**. And whereas the **Bonferonni method** divides α by k to get a per comparison (contrast) type I error level of α/k. All methods have relatively good power – with orthogonal contrasts being ranked first, Bonferroni or Tukey's HSD being ranked second, and Scheffe's interval being ranked third – and can be applied to equal or unequal group sizes (Gill 1977). Additional information about the Bonferroni method, Tukey's HSD method, and Scheffe's interval method follows:

- **The Bonferron**i method is general, simple to use, and widely applicable. However, since the experiment wise error rate is divided by the number of comparisons to get the per comparison error rate, its power may be reduced. The type I error may be smaller than needed. For example, taking into account the correlation among the comparisons, would lead to larger per comparison type I error rates;

- **Tukey's honest significant difference** method is best for all-possible pairwise comparisons;

- **Scheffé's** method is good for all possible contrasts (a contrast is a linear combination of contrast constants and group means such that the sum of the constants is zero) including unplanned contrasts among sets of means other than just pairwise comparisons. It can be used to control the overall confidence level or type I error level.

4.3.2 Analysis of Data in SAS and R

We use the tumor size versus smoking data to demonstrate these models through SAS and R statistical software. Outputs from those statistical packages are at times truncated to avoid redundancy. The equation part of the model is $Y_{ij} = \mu + \alpha_i + \varepsilon_{ij}$.

4.3.2.1 Analysis of Data with SAS Program

```
proc glm data=tumorsize PLOT(MAXPOINTS=NONE);
class SmokingHx ;
model tumorsize=SmokingHx ;
lsmeans SmokingHx /adjust=Bon cl;
*lsmeans SmokingHx /adjust=tukey cl; * we did not
include this in our output as the results are similar;
*lsmeans SmokingHx /adjust=scheffe cl; * we did not
include this in our output as the results are similar;
run;
```

This SAS code is responsible for running $Y_{ij} = \mu + \alpha_i + \varepsilon_{ij}$
The SAS output is attached here.

The GLM Procedure
Dependent Variable: tumorsize

Source	DF	Sum of Squares	Mean Square	F Value	Pr > F
Model	2	397568.591	198784.295	2007.38	<.0001
Error	8522	843905.786	99.027		
Corrected Total	8524	1241474.377			

> The F test statistic value is 2007.38 with p-value of <.0001. This tells us that the mean tumor sizes for the three groups are not the same. We assume that the variance is the same for the three groups. That common variance or the model variance is estimated to be 99.027.

The GLM Procedure
Least Squares Means: Adjustment for Multiple Comparisons: Bonferroni

SmokingHx	Tumorsize LSMEAN	LSMEAN Number
current	83.7734099	1
former	72.0220827	2
never	66.2026187	3

The mean tumor size for current smokers is 83.773; for former smokers, it is 72.022; and for those who never smoked, it is 66.202. We used the Bonferroni method to address the multiple comparisons for the three groups. The three groups provided 3*(3–1)/2 = 3 pairwise comparisons.

Least Squares Means for effect SmokingHx		
Pr > |t| for H0: LSMean(i)=LSMean(j)		
Dependent Variable: tumorsize		
i/j	1	2	3
1		<.0001	<.0001
2	<.0001		<.0001
3	<.0001	<.0001	

The three comparisons showed that they were all significantly different. We made this conclusion as the p-value (<.0001) for each comparison is less than $\alpha = \dfrac{0.05}{3} = 0.0167$.

Note to SAS readers: We obtained similar results with SAS commands.

```
lsmeans SmokingHx /adjust=tukey cl;
lsmeans SmokingHx /adjust=scheffe cl;
```

We did not include this in our output, as the results are similar.

4.3.2.2 Analysis of Data with R Program

Analysis of Variance (ANOVA) and multiple comparisons for continuous data.

```
Model = aov(tumorsize~SmokingHx,data=tumorsize) ## ANOVA
                                                      Model
summary(model)
##              Df    Sum Sq Mean Sq  F value   Pr(>F)
## SmokingHx    2    397569  198784   2007     <2e-16 ***
## Residuals  8522    843906  99
```

> The F-test statistic value is 2007 with p-value of <2e-16. This tells us that
> the mean tumor sizes for the three groups are not the same.

```
pairwise.t.test(tumorsize$tumorsize,SmokingHx,
                "bonferroni") ## Bonferroni
##Pairwise comparisons using t tests with pooled SD
##data: tumorsize$tumorsize and SmokingHx
##        current  former
## former <2e-16   -
## never  <2e-16   <2e-16
```

> We used the Bonferroni method to obtain the comparison Type I error
> rate that addresses the multiple comparisons for the three groups. The
> three groups provided 3*(3–1)/2 = 3 pairwise comparisons, p <2e-16.

```
## P value adjustment method: bonferroni
TukeyHSD(model)              ## Tukey
##                Tukey multiple comparisons of means
## 95% family-wise confidence level
## Fit: aov(formula = tumorsize ~ SmokingHx, data = tumorsize)
## $SmokingHx
##                   diff          lwr                 upr        p adj
## former-current -11.751327   -12.550254   -10.952400    0
## never-current  -17.570791   -18.223113   -16.918470    0
## never-former    -5.819464    -6.471785    -5.167143    0
library(agricolae)
## alpha = 0.01
scheffe1 = scheffe.test(model,"SmokingHx",group=FALSE,alpha = 0.01)
scheffe1
## Scheffe
## $statistics
##              Mean       CV          MSerror    CriticalDifference
##           70.88067   14.0394      99.02673    0.9124561
## $parameters
##              Df       ntr     F        Scheffe   alpha     test      name.t
##            8522       3     4.60766   3.035674   0.01      Scheffe   SmokingHx
## $means
##              tumorsize     std         r        Min        Max
## current     83.77341    9.992957    1705    53.38128    116.4579
## former      72.02208   10.033507    1705    40.67090    104.5832
## never       66.20262    9.909661    5115    33.96859    102.9499
## $comparison
##                   Difference    pvalue sig   LCL         UCL
## current - former   11.751327     0 ***      10.911535    12.591119
## current - never    17.570791     0 ***      16.885104    18.256478
## former - never      5.819464     0 ***       5.133777     6.505151
## $groups
## NULL
## alpha= 0.05
scheffe2 = scheffe.test(model,"SmokingHx",group=FALSE)
scheffe2
## $statistics
##              Mean       CV          MSerror    CriticalDifference
##           70.88067   14.0394      99.02673    0.7358675
## $parameters
##              Df           ntr       F Scheffe    alpha     test    name.t
```

```
##                 8522        3        2.996786        2.448177  0.05      Scheffe      SmokingHx
## $means
##                  tumorsize       std        r           Min          Max
## current         83.77341     9.992957     1705        53.38128     116.4579
## former          72.02208    10.033507     1705        40.67090     104.5832
## never           66.20262     9.909661     5115        33.96859     102.9499
## $comparison
##                  Difference  pvalue sig      LCL          UCL
## current - former  11.751327     0 ***      10.997164     12.505491
## current - never   17.570791     0 ***      16.955019     18.186563
## former - never     5.819464     0 ***       5.203692      6.435236
## $groups
## NULL
##Two-way ANOVA
model2 = aov(tumorsize~SmokingHx+CancerStage,data=tumorsize)
summary(model2)
##               Df       Sum Sq      Mean Sq    F value    Pr(>F)
## SmokingHx      2       397569      198784     2098.4    <2e-16 ***
## CancerStage    3        36890       12297      129.8    <2e-16 ***
## Residuals   8519       807016         95
model3 = aov(tumorsize~SmokingHx+CancerStage+SmokingHx*CancerStage,
             data=tumorsize)                 ##Two-way ANOVA with interaction
summary(model3)
##                        Df     Sum Sq    Mean Sq    F value Pr(>F)
## SmokingHx               2     397569    198784     2099.544   <2e-16 ***
## CancerStage             3      36890     12297      129.876   <2e-16 ***
## SmokingHx:CancerStage   6       1007       168        1.773   0.1
## Residuals            8513     806009        95
```

```
pairwise.t.test(tumorsize$tumorsize,CancerStage,
"bonferroni")
## Pairwise comparisons using t tests with pooled SD
## data: tumorsize$tumorsize and CancerStage
##        I          II         III
## II    < 2e-16    -          -
## III   < 2e-16    0.004      -
## IV    8.8e-07    1.000      0.012
## P value adjustment method: bonferroni
library(lsmeans)
## The 'lsmeans' package is being deprecated.
## Users are encouraged to switch to 'emmeans'.
## See help('transition') for more information,
                               including how
## to convert 'lsmeans' objects and scripts to work with
                    'emmeans'.
SC = pairs(lsmeans(model3,~ CancerStage|SmokingHx))       ##
            Bonferroni
test(SC,adjust="bonferroni")
```

We use the Bonferroni method to obtain the multiple comparisons for the three groups. The three groups provided 3*(3–1)/2 comparisons, p <2e-16.

> 95% CI for the difference between group means is constructed by making adjustment for the inflated type I error due to the many comparisons (1 v 2, 1 v 3, 2 v3).

4.3.3 Continuous Response with a Categorical Factor: Nonparametric Tests

When we have three or more groups to compare and wish to do so without assuming the outcomes came from a normally distributed population, we can use the Kruskal-Wallis nonparametric test. The test utilizes the ranks of the observations rather than the magnitude of the observations. The assumptions necessary to use Kruskal-Wallis are:

1. There are three or more groups to compare;
2. Each group of participants is exposed to different conditions, but within each group, the outcomes are independent.

In short, use the Kruskal-Wallis test if the data are not normally distributed; if the variances for the different conditions are markedly different; or if the data are measurements on an ordinal scale. Rejection of the null hypothesis by the Kruskal-Wallis test tells us that the differences between the groups are so large that they are unlikely to have occurred by chance. It enables one to reject the null hypothesis that the means or medians of the groups are the same. We have to utilize multiple range tests or post-hoc tests to identify the groups whose means or medians are different.

4.3.3.1 Analysis of Data with SAS Program

```
***Kruskal-Wallis;
proc nparlway data=tumorsize wilcoxon;
class SmokingHx;
var tumorsize;
run;
```

The NPAR1WAY Procedure

		Wilcoxon Scores (Rank Sums) for Variable tumorsize			
		Classified by Variable SmokingHx			
SmokingHx	N	Sum of Scores	Expected Under H0	Std Dev Under H0	Mean Score
former	1705	7809898.0	7268415.0	90894.252	4580.58534
never	5115	16964763.0	21805245.0	111322.269	3316.66921
current	1705	11567414.0	7268415.0	90894.252	6784.40704

Kruskal-Wallis Test	
Chi-square	2574.239
DF	2
Pr > Chi-square	<.0001

Use the Kruskal-Wallis test the same way we use the ANOVA F-test to compare the group means. The test has a value of 2574.2378 with a p-value <.0001, showing the group means are not all the same. The box plots in Figure 4.2 provide a comparison of the group means.

4.3.3.2 Analysis of Data with R Program

```
Kruskal Wallis
kruskal.test(tumorsize~SmokingHx,data=tumorsize)
## Kruskal-Wallis rank sum test
## data: tumorsize by SmokingHx
## Kruskal-Wallis chi-squared = 2574.2, df = 2, p-value
                               < 2.2e-16
```

Use the Kruskal-Wallis test the same way we use the ANOVA F test to compare the group means. The test has a value of 2574.2 with a p-value < 2.2e-16, showing the group means are not all the same.

4.4 Modeling Binary Response with One Categorical Factor

4.4.1 Models

In this section, modeling of a binary response variable with a categorical factor is addressed. The data may be displayed as an rxc (r, rows by c, columns) contingency table. For example, we may be interested in using the HDP dataset to assess whether there is any association between remission of lung cancer and smoking history. The data may be summarized in a rectangular display of 3 rows by 2 columns as shown in Table 4.3.

4.4.1.1 Test of Homogeneity-Hypotheses

We wish to know if the probabilities are the same across several subpopulations. In particular, we want to know if the probability of cancer remission

TABLE 4.3

Classification of Smoking History by Remission

Smoking History	No Remission [0]	Remission [1]	Total
Current	1140	565	1705
Former	1190	515	1705
Never	3674	1441	5115
Total	6004	2521	8525

is the same across all groups (current, former, never smoker). This is a test based on the observed proportions. The null hypothesis is that the probabilities or true proportions of cancer remission are the same for all three groups. The question is still about the means across groups, but since the response on each patient is binary, the mean is the probability or proportion.

In the HDP data about *Tumor Size*, there are 8525 patients of which 1705 individuals make up the current smoker group, 1705 individuals make up the former smoker group, and 5115 individuals make up the never smoker group. Assume that outcomes (remission, non-remission) are independent. Whether an individual has remission or not may be regarded as a Bernoulli trial. Each of the three groups provides a separate binomial distribution (number in cancer remission among the total in each group).

4.4.1.2 Test of Independence-Hypotheses

One can also ask, if smoking history is related to the remission of cancer. This is the test of independence. The test statistic and the p-value are the same for both tests (test of homogeneity and test of independence), but the interpretations are different. In assessing the test of independence in the HDP Tumor Size data, consider it as 8525 patients or 8525 independent trials. The outcomes are binary (cancer in remission or cancer not in remission) crossed with smoking history (current, former, and never). Both variables are considered as response variables. Each trial has two sets of responses resulting in one of six possible outcomes (remission (2) by smoking history (3)). Thus, there are multivariate responses. There is a fixed overall total. Therefore, instead of a binomial distribution (two possible outcomes), there is a multinomial distribution (more than two possible non-continuous outcomes), and the test is the test of independence.

4.4.1.3 Pearson Chi-Square

Consider a distance measure X^2 construct that takes the key information about the sample (cell count, row totals, column totals) into a single value. This is a test statistic that measures distance between the data values and the values expected under the null hypothesis H_0. Based on the size of the test statistic, a decision about the null hypothesis is made. In this case, we refer to that overall measure as a Pearson chi-square test statistic. The model under the test of independence says the cell probability equates to the product of the row probability and column probability,

Cell probability = *Row* probability times *column* probability

$$p_{ij} = p_{i+}p_{+j}$$

where p_{ij} (*cell probability*) is the probability that the outcome is in row i and column j, p_{i+} (*row probability*) is the probability in row i and p_{+j} (column probability) is the probability in column j. Under H_0, it assumes independence; the distance measure X^2 (Pearson chi-square test) follows a chi-square distribution with (r-1)(c-1) degrees of freedom, where r is the number of rows and c is the number of columns of the table.

$$X^2 = \Sigma\Sigma \left(\frac{Obs - Exp}{\sqrt{(Exp)}} \right)^2$$

where the summation is over all cells, Obs is the cell count observed, Exp is the expected cell value obtained by taking the cell probability times the total observed. There are other chi-square test statistics, which can be used to test the null hypothesis of independence of the row and column categories.

Decision: We obtain the value of the *test statistic* and use it to decide if we have sufficient evidence to support the research hypothesis (and doubt the present state of the system). We rely on the distribution of the test statistic. In this case, we rely on a chi-square distribution with appropriate degrees of freedom as shown in Figure 4.3. We refer to the p-value of that distribution. We obtain the p-value based on the tail of the distribution beyond the test statistic value. We need a cut point in the tail of the distribution to determine if it is small enough to make a decision to claim that the research statement is the one to support. We refer to the cut point on the graph as the critical value. The area under the distribution beyond the critical value is the significance level or α level. This level is usually set by the client or the researcher in advance of conducting the experiment.

Many statisticians believe that we should report the p-value rather than compare it with the probability of type I or the risk level. A rule of thumb is if p-value < α, we have sufficient evidence to say that smoking behavior is related to the remission in cancer. The p-value is a conditional probability (i.e., the probability of the event producing the observed data or events more extreme conditional on the null hypothesis being true). The smaller its value, the more we are inclined to believe our research (alternative) hypothesis. In a recent report, the American Statistical Association has outlined some guidelines in the use and misuse of p-values (Wasserstein and Lazar 2016).

Comment: We want to know if smoking history had an impact on the outcome (remission or no remission of cancer). The Pearson chi-square test, as well as the other chi-square tests, is non-directional (association or no association). For now, we concentrate on one factor for the binary outcome. In Chapter 8, we use the binary logistic regression model, which allows us to have multiple factors for the binary outcomes.

4.4.2 Analysis of Data using SAS and R

We outline chi-square test for the rxc contingency table and discuss their output from different statistical programs in SAS and R.

4.4.2.1 Analysis of Data with SAS Program

```
proc freq data=tumorsize;
tables remission*SmokingHx/all;
run ;
```

Statistics for Table of remission by SmokingHx			
Statistic	DF	Value	Prob
Chi-Square	2	15.5510	0.0004
Likelihood Ratio Chi-Square	2	15.3703	0.0005
Mantel-Haenszel chi-square	1	15.4250	<.0001

The test of independence or the test of homogeneity has as statistic values Pearson chi-square = 15.551, with p = 0.0004; likelihood ratio chi-square = 15.37, with p = 0.0005. Both the test of independence and the test of homogeneity are significant. Thus from these results, we conclude that there is a relationship between smoking and remission (test of independence), or we conclude that the probability of remission differs across the three groups (test of homogeneity).

Mantel-Haenszel chi-square test of our 3x2 contingency table assumes that the margins of the table are fixed. It tests the null hypothesis that the probability of remission is the same among all three groups, or whether there is any association between smoking and recurrence of cancer. The value of the test statistic = 15.425, with p<.0001, indicating that there is a statistically significant negative relationship between smoking and remission.

Phi Coefficient		0.0427
Contingency Coefficient		0.0427
Cramer's V		0.0427
Statistic	**Value**	**ASE**
Gamma	−0.0810	0.0211
Kendall's Tau-b	−0.0396	0.0105

The table was truncated, but there are several other measures of association available in SAS. In this case we have a gamma measure or −0.0810 with asymptotic standard error (ASE) 0.0211. The ratio is −0.0810/0.0211 = −3.84 (which gives a p-value <0.01).

4.4.2.2 Analysis of Data with R Program

Contingency tables and chi-square tests for categorical data.

```
library(gmodels)
freq1 = CrossTable(tumorsize$remission, tumorsize$SmokingHx,
        expected = TRUE, prop.r=TRUE, prop.c=TRUE,
        prop.t=TRUE, prop.chisq=TRUE, chisq = TRUE)
```

```
##                  Cell Contents
## |                       N |
## |              Expected N |
## |    Chi-square contribution |
## |            N / Row Total |
## |            N / Col Total |
## |          N / Table Total |
## Total Observations in Table: 8525
##| tumorsize$SmokingHx
## tumorsize$remission | current |former  | never  | Row Total|
##-------------------- |---------- |------------ |---- |
##          0 |     1140|     1190|       3674|  6004|
##            | 1200.800| 1200.800|   3602.400|      |
##            |    3.078|    0.097|      1.423|      |
##            |    0.190|    0.198|      0.612| 0.704|
##            |    0.669|    0.698|      0.718|      |
##            |    0.134|    0.140|      0.431|      |
## ------------- |-------- |--------- |-------------|------|
##          1|      565|      515|       1441|  2521|
##            |  504.200|  504.200|   1512.600|      |
##            |    7.332|    0.231|      3.389|      |
##            |    0.224|    0.204|      0.572| 0.296|
##            |    0.331|    0.302|      0.282|      |
##            |    0.066|    0.060|      0.169|      |
## ------------- |-------- |--------- |-------------|------|
## Column Total |     1705|     1705|       5115|  8525|
##            |    0.200|    0.200|      0.600|      |
## ------------- |-------- |--------- |-------------|------|
```

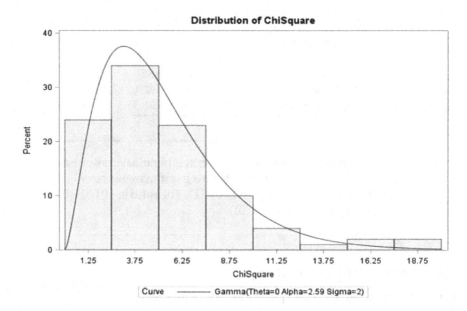

FIGURE 4.3
Distribution of a chi-square random variable

```
## Statistics for All Table Factors
## Pearson's Chi-squared test
## Chi^2 = 15.55098  d.f. = 2  p = 0.0004199018
```

The test of independence or the test of homogeneity has as statistic values Pearson chi-square = 15.55098, with p = 0.0004199018; for both test of independence and test of homogeneity, they are statistically significant. Thus, from these results, we conclude that there is a relationship between smoking and remission (test of independence), or we conclude that the probability of remission differs across the three groups (test of homogeneity).

4.5 Modeling of Continuous Response with Two Categorical Factors

4.5.1 Models

Thus far, we have considered the impact of one factor on the mean of a distribution from the exponential family of distributions. Such models may be acceptable in the design of experiments where researchers have control of factors. However, when we deal with observational studies, the one-factor model will probably not be appropriate, as there may be a cadre of factors influencing the output system. In this section, we expand on one factor and examine two categorical factors on a continuous response. *We are focusing here on the type of variable setup and not necessarily on the practicality of the science.*

Consider a two-way ANOVA model. In particular, examine the simultaneous effect of smoking and stages of cancer on tumor size. $Y_{ijk} = y_{ijk}$ for the ith person in the jth smoking group and the kth cancer stage and postulate that y_{ijk} may be influenced by smoking history and stages of cancer.

To test whether the means of continuous outcomes are statistically different, we fit a two-way analysis of variance (ANOVA) model. It is limited in that it will only tell if the means among the groups are equal or not. The two-way ANOVA model compares the means of the smoking groups and means of the stages of cancer through the partitioning of the variance of the outcomes. Thus, for each cell with multiple measures is demonstrated in Table 4.4. There are 3 rows and 3 columns each with two observations in each cell.

Two-way ANOVA model: The two-way ANOVA statistical model is

$$Y_{ijk} = \mu + \alpha_i + \beta_j + \varepsilon_{ijk}$$

where the i indexes the rows, α_i is the effect of the ith row mean in comparison to the overall mean, j indexes the columns, β_j is the effect of the jth column mean in comparison to the overall mean, ε_{ijk} represents random measurement error, and k represents the number of replications in each cell. The statistical model

$$y_{ijk} = \mu_{ij} + \varepsilon_{ijk}$$

TABLE 4.4

Two-Way ANOVA Model: Setup for Demonstration

	Col 1	Col 2	Col 3	
Row 1	$Y_{111}=45$	$Y_{121}=40$	$Y_{131}=52$	$\bar{Y}_{1**}=50$
	$Y_{112}=55$	$Y_{122}=60$	$Y_{132}=48$	
	$\bar{Y}_{11*}=50$	$\bar{Y}_{12*}=50$	$\bar{Y}_{13*}=50$	
Row 2	$Y_{211}=35$	$Y_{221}=30$	$Y_{231}=21$	$\bar{Y}_{2**}=29.33$
	$Y_{212}=35$	$Y_{222}=32$	$Y_{232}=23$	
	$\bar{Y}_{21*}=35$	$\bar{Y}_{22*}=31$	$\bar{Y}_{23*}=22$	
Row 3	$Y_{311}=15$	$Y_{321}=10$	$Y_{331}=11$	$\bar{Y}_{3**}=13.33$
	$Y_{312}=19$	$Y_{322}=12$	$Y_{332}=13$	
	$\bar{Y}_{31*}=17$	$\bar{Y}_{32*}=11$	$\bar{Y}_{33*}=12$	
	$\bar{Y}_{*1*}=34$	$\bar{Y}_{*31*}=30.67$	$\bar{Y}_{*31*}=28$	$\bar{Y}_{***}=33.89$

where $\mu_{ij} = \mu + \alpha_i + \beta_j =$ is the mean of the ij^{th} cell, μ is the overall mean across all cells, and ε_{ijk} is a measure of the unexplained variation for the ij^{th} cell mean from the observation, y_{ijk}. Assume that random variable ε_{ijk} is distributed normally with mean zero and variance σ^2, which indicates that observations y_{ijk} come from a normal distribution with mean μ_{ij} and variance σ^2.

Two-way ANOVA-F-statistic for hypotheses: To answer the research question, construct a test statistic that utilizes key information (cell means, group means, overall group means, sample size, standard error) from the sample of patients. Based on the size of the value of the test statistic, make a decision (reject or not) about the null hypothesis. In particular, our test statistic is the two-way ANOVA F-statistic. This is a two-way ANOVA since there are two categorical variables as factors, such as smoking history and cancer stage. The ANOVA F-test for the row factor enables the analyst to determine if there are significant differences among row means but does not say which rows are significantly different from one to another. In fact, the model is testing

$H_{10} : \mu_{cur} = \mu_{for} = \mu_{nev}$ versus H_{11}: not all equal,
and
$H_{20} : \mu_{stage1} = \mu_{stage2} = \mu_{stage3} = \mu_{stage4}$ versus H_{21}: not all equal.

4.5.2 Analysis of Data with SAS Program

```
*** Two-Way ANOVA;
proc glm data=tumorsize PLOT(MAXPOINTS=NONE);
class SmokingHx CancerStage ;
model tumorsize=SmokingHx CancerStage;
lsmeans SmokingHx CancerStage/adjust=Bon cl;
lsmeans SmokingHx CancerStage/adjust=tukey cl;
lsmeans SmokingHx CancerStage/adjust=scheffe cl;
run;
```

> We only included the output for the Bon = Bonferroni method, but the results are similar for the Tukey and Scheffe methods.

The GLM Procedure

Class Level Information		
Class	Levels	Values
SmokingHx	3	current former never
CancerStage	4	I II III IV

The GLM Procedure
Dependent Variable: tumorsize

Source	DF	Sum of Squares	Mean Square	F Value	Pr > F
Model	5	434458.386	86891.677	917.24	<.0001
Error	8519	807015.990	94.731		
Corrected Total	8524	1241474.377			

The F-value of 917.24 with p <.0001 informs us that $H_0 : \mu_{cur} = \mu_{for} = \mu_{nev}$ and $H_0 : \mu_{stage1} = \mu_{stage2} = \mu_{stage3} = \mu_{stage4}$ are rejected. There are differences in tumor sizes among the smoking groups and or differences among the cancer stages.

Source	DF	Type III SS	Mean Square	F Value	Pr > F
SmokingHx	2	414886.2818	207443.1409	2189.81	<.0001
CancerStage	3	36889.7958	12296.5986	129.81	<.0001

The F-value of 2189.81 with p <.0001 informs us that $H_0 : \mu_{cur} = \mu_{for} = \mu_{nev}$ is false, and F-value 129.81 with p <.0001 tells us that $H_0 : \mu_{stage1} = \mu_{stage2} = \mu_{stage3} = \mu_{stage4}$ is false. There are differences in tumor sizes among the smoking groups and among the cancer stages.

Least Squares Means for effect SmokingHx

Pr > |t| for H0:
LSMean(i)=LSMean(j)

Dependent Variable: tumorsize

i/j	1	2	3
1		<.0001	<.0001
2	<.0001		<.0001
3	<.0001	<.0001	

The multiple comparisons among the group means are significantly different in terms of tumor sizes, p <.0001. The comparison of group 1 to group 2, and group 2 to group 3, as well as group 1 to group 3 are significantly different.

Least Squares Means for Effect SmokingHx				
i	j	Difference Between Means	Simultaneous 95% Confidence Limits for LSMean(i)-LSMean(j)	
1	2	12.579410	11.770142	13.388679
1	3	20.154921	19.424651	20.885191
2	3	7.575511	6.888152	8.262869

The difference in means between groups 1 and 2 is 12.57, with a range of values from 11.77 to 13.39. The difference in means between groups 1 and 3 is 20.15, with a range of values from 119.42 to 20.89.
The difference in means between groups 2 and 3 is 7.58, with range of values from 6.89 to 8.26.

	Least Squares Means for effect CancerStage			
	Pr > |t| for H0: LSMean(i)=LSMean(j)			
	Dependent Variable: tumorsize			
i/j	1	2	3	4
1		<.0001	<.0001	<.0001
2	<.0001		<.0001	<.0001
3	<.0001	<.0001		<.0001
4	<.0001	<.0001	<.0001	

Question: The multiple comparisons among smoking group means in terms of differential tumor sizes are significant, p <.0001. What does this mean? Explain.

4.5.3 Analysis of Data with R Program

Analysis of Variance (ANOVA) and multiple comparisons for continuous data.

```
##Two-way ANOVA
model2 = aov(tumorsize~SmokingHx+CancerStage,data=tumorsize)
summary(model2)
##              Df     Sum Sq    Mean Sq  F value  Pr(>F)
## SmokingHx    2      397569    198784   2098.4   <2e-16 ***
```

```
## CancerStage  3      36890    12297   129.8   <2e-16 ***
## Residuals    8519   807016   95
## ---
## Signif. codes:    0 '***' 0.001 '**' 0.01 '*' 0.05 '.' 0.1 ' ' 1
```

> The F-value of 2098.4 with p <2e-16 informs us that $H_0 : \mu_{cur} = \mu_{for} = \mu_{nev}$ is rejected, and F-value 129.81 with p <2e-16 informs us that $H_0 : \mu_{stage1} = \mu_{stage2} = \mu_{stage3} = \mu_{stage4}$ is rejected. There are differences in tumor sizes among the smoking groups and among the cancer stages. To determine where the differences may exist one can use a multiple comparison test.

4.6 Modeling of Continuous Response: Two-Way ANOVA with Interactions

The two-way ANOVA model with interaction is fitted. Interaction means that patterns of a factor change as we move across the levels of the other factor or covariate. For example, the interpretation of interaction, in our defense, means that the differences in the response means for never smoked to former smoker to current smoker are not the same among all the differences from stage 1, to stage 2, to stage 3, and to stage 4. **To explore:** Outline the two-way ANOVA model with interaction by adding on to the previous two-way model of smoking history and cancer stages. We discuss the output from two different statistical programs, SAS and R.

4.6.1 Analysis of Data with SAS Program

*** With Interaction;
```
ods graphics on;
proc glm data=tumorsize PLOT(MAXPOINTS=NONE) ;
class SmokingHx CancerStage;
model tumorsize=SmokingHx CancerStage SmokingHx*CancerStage;
run;
```

Dependent Variable: tumorsize

Source	DF	Sum of Squares	Mean Square	F Value	Pr > F
Model	11	435465.537	39587.776	418.12	<.0001
Error	8513	806008.840	94.680		
Corrected Total	8524	1241474.377			

Source	DF	Type III SS	Mean Square	F Value	Pr > F
SmokingHx	2	51851.15571	25925.57786	273.82	<.0001
CancerStage	3	18941.27534	6313.75845	66.69	<.0001
SmokingHx*CancerStag	6	1007.15027	167.85838	1.77	0.1004

> The F-value of 273.82 with p <.0001 informs us that $H_0 : \mu_{cur} = \mu_{for} = \mu_{nev}$ is rejected, and F-value 66.69 with p <.0001 tells us that $H_0 : \mu_{stage1} = \mu_{stage2} = \mu_{stage3} = \mu_{stage4}$ is rejected. The F-value of 1.77 with p =0.1004 informs us that there is no statistically significant interaction (at the 0.05 level).
>
> There are differences in tumor sizes among the smoking groups and among the cancer stages. The interaction (smoking history by cancer stage) is not significant. Thus, the difference in smoking history remains similar in its distribution at all the cancer stage levels.

4.6.2 Analysis of Data with R Program

```
##Two-way ANOVA with interaction
model3 = aov(tumorsize~SmokingHx+CancerStage+SmokingHx
*CancerStage,
    data=tumorsize)
summary(model3)
```

```
##                      Df    Sum Sq  Mean Sq  F value   Pr(>F)
## SmokingHx             2    397569  198784   2099.544  <2e-16 ***
## CancerStage           3    36890   12297    129.876   <2e-16 ***
## SmokingHx:CancerStage 6    1007    168      1.773     0.1
## Residuals             8513  806009  95
```

```
SC = pairs(lsmeans(model3,~ CancerStage|SmokingHx))
## Bonferroni for interaction
test(SC,adjust="bonferroni")
```

```
## SmokingHx = current:
##  contrast   estimate    SE          df      t.ratio    p.value
##I - II       -3.503988   0.5167994   8513    -6.780     <.0001
##I - III      -5.310563   1.0620078   8513    -5.000     <.0001
##I - IV       -9.663102   4.8741527   8513    -1.983     0.2847
##II - III     -1.806575   1.1045791   8513    -1.636     0.6118
##II - IV      -6.159113   4.8836052   8513    -1.261     1.0000
##III - IV     -4.352539   4.9709535   8513    -0.876     1.0000
##SmokingHx = former:
##contrast       estimate    SE          df      t.ratio p.value
```

```
##I - II          -3.242231   0.5050229   8513    -6.420     <.0001
##I - III         -5.807991   0.8084369   8513    -7.184     <.0001
##I - IV         -11.349541   1.6847813   8513    -6.737     <.0001
##II - III        -2.565760   0.8011959   8513    -3.202     0.0082
##II - IV         -8.107311   1.6813187   8513    -4.822     <.0001
##III - IV        -5.541550   1.7959272   8513    -3.086     0.0122
## SmokingHx = never:
##contrast         estimate   SE          df t.ratio p.value
##I - II          -2.303602   0.4108093   8513    -5.607     <.0001
##I - III         -4.003127   0.4357540   8513    -9.187     <.0001
##I - IV          -7.004115   0.4899782   8513 -14.295       <.0001
##II - III        -1.699525   0.3333422   8513    -5.098     <.0001
##II - IV         -4.700513   0.4016393   8513 -11.703       <.0001
##III - IV        -3.000987   0.4271198   8513    -7.026     <.0001
## P value adjustment: bonferroni method for 6 tests
SC2= pairs(lsmeans(model3,~ CancerStage+SmokingHx
         +CancerStage*SmokingHx))
test(SC2,adjust="bonferroni")
##contrast                        estimate    SE          df    t.ratio p.value
##I,current - II,current      -3.5039885   0.5167994 8513   -6.780<.0001
##I,current - III,current     -5.3105631   1.0620078 8513   -5.000<.0001
##I,current - IV,current      -9.6631018   4.8741527 8513   -1.9831.0000
```

The F-value of 2099.544 with p <.0001 informs us that $H_0 : \mu_{cur} = \mu_{for} = \mu_{nev}$ is rejected, and F-value 129.876 with p <.0001 tells us that $H_0 : \mu_{stage 1} = \mu_{stage 2} = \mu_{stage 3} = \mu_{stage 4}$ is rejected. The F-value of 1.77 with p =0.1004 informs us that there is no statistically significant interaction at the 0.05 level.

There are differences in tumor sizes among the smoking groups and among the cancer stages. The interaction (smoking history by cancer stage) is not significant. Thus, the differences in smoking history are similar across all the cancer stage levels.

4.7 Summary and Discussion

The sampling distributions of Pearson chi-square test and the G^2 likelihood ratio chi-square test are based on large samples; i.e., they are asymptotic procedures. They rely on large amount of observations.

However, when the dataset is small, these asymptotic methods may fail to produce reliable results. In such cases, it is advisable to rely on the exact distribution of the test statistic. By so doing, one can obtain an accurate p-value without relying on assumptions that may not be valid in the data.

The exact computations always produce a reliable result, regardless of the size, distribution, sparseness, or balance of the data. It is a fact that calculating

TABLE 4.5

Guide to Selecting Appropriate Models

Covariate or Factor	Variable of Interest	Distribution	Test
N/A	continuous	Normal	one sample t-test
N/A	continuous	None	Wilcoxon rank signed test
N/A	binary	Binomial	Z test of proportions
N/A	binary	Binomial	exact Binomial test
N/A	two categorical	Multinomial	test of independence/ Pearson chi-square test
Categorical	categorical	Multinomial	test of homogeneity/ Pearson chi-square test
Binary	continuous	Normal	two-sample independent t-test
Binary	continuous	None	Mann Whitney/Wilcoxon
Binary	binary	Binomial	Pearson chi-square test
Categorical	categorical	Binomial	Fisher's exact test
Categorical	continuous	Normal	one-way ANOVA
Categorical	continuous	None	Kruskal-Wallis
Categorical	categorical	product binomial	test of homogeneity/ Pearson chi-square test
Two categorical	continuous	Normal	two-way ANOVA

exact results can be computationally intensive and time-consuming and can sometimes exceed the memory limits of your machine. However, programs like R and C++ are great options (Troxler, Lalonde, and Wilson 2011).

All the models presented thus far had either no explanatory variable or covariate, or one continuous variable, or one categorical variable. These models are summarized in Table 4.5. The sampling units in these models are measured once. Therefore, the sampling unit is the same as the observational unit. Hence, that the observations are independent means that the sampling units are independent. Thus, in the discussions in earlier chapters, we do not distinguish between sampling units and observational units. However, in cases where the sampling units are measured more than once, the observations are correlated (not independent), but the sampling units are still independent.

A summary of models stating their type of factor or covariate and the type of variable of interest is given in Table 4.5. In Table 4.5, we have the input variable (covariate) as categorical (binary as a special case). The variable of interest (output) is binary, categorical, or continuous. There is one variable of interest discussed in these models, except for test of independence with two categorical variables as response. In that unique case, we have two categorical (binary) responses and no covariate.

4.8 Exercises

Use the tumor size data to answer the following questions.

1. Do gender and family history impact BMI?
   a. Is there a differential effect?
   b. Is there a family history effect?
2. Is there a relationship between cancer stage and gender?
   a. How would you analyze if the cells had structural zeros?
   b. How would you analyze if the cells had random zeros?
   c. How would you analyze if the cells were sparse?
3. How would you analyze if you were interested in knowing if smokers were more likely to have remission than nonsmokers were?

References

Day, R.W., Quinn G.P.: Comparisons of treatments after an analysis of variance in ecology. *Ecological Monography*, 59, 433–463 (1989)

Gill, J.: *Design and Analysis of Experiments*. The Iowa State University Press, Ames. IA, Vol. 1, pp. 185–186 (1977)

Troxler, S., Lalonde, T, Wilson, J.R.: Exact logistic models for nested binary data. *Statistics in Medicine*, 30(8), 866–876 (2011)

Wasserstein, R.L., Lazar, N.A.: The ASA statement on *p*-Values: Context, process, and purpose. The American Statistician, 70(2), 129–133 (2016). https://doi.org/10.1080/00031305.2016.1154108

5

Statistical Modeling of Continuous Outcomes with Continuous Explanatory Factors: Linear Regression Models

5.1 Research Interest/Question

As in previous chapters, this chapter concentrates on modeling the mean of a continuous random variable, but with the aid of multiple continuous explanatory factors or covariates. Initially, the interest is in modeling the impact of one continuous covariate on one continuous response. This extends to examine the impact of several continuous covariates simultaneously on the continuous response or a mixture of continuous and categorical covariates simultaneously on the continuous response. Typically, in statistical terminology, one continuous variable on one continuous response is referred to as *simple linear regression*. A set of continuous variables on a continuous response is referred to as *multiple linear regression*. In particular, this chapter reports first the variable age on body mass index (BMI); then age, waist size, and albumin on BMI; and finally, age, waist size, albumin, and race on BMI. A stepwise procedure to aid in selecting significant covariates in model building is presented.

The fit of our models uses the National Health and Nutrition Examination Survey (NHANES) data to demonstrate. The NHANES is unique as it combines personal interviews with standardized physical examinations and laboratory tests to gather information about illness and disability. The interviews are conducted in the home and at a mobile examination center. Approximately 5000 people per year participate in NHANES as a representation of the entire US population. A subset of these data is presented in Table 5.1. This subset contains ID (i.e., SEQN-respondent sequence number), BMI, gender, taking insulin or diabetes meds (TX), waist size (measured in

DOI: 10.1201/9781003315674-5

TABLE 5.1

A Subset of the NHANES Dataset

SEQN	Gender	Tx	BMI	Age	Obese
31106	male	0	15.25	22.00	1
31107	male	0	29.92	72.00	0
31108	female	1	22.16	26.00	1
31109	male	0	22.60	14.00	1
31110	female	0	29.41	34.00	1
31111	male	0	25.18	13.00	1
31112	male	1	17.63	12.00	1
31113	male	0	21.28	55.00	1
311114	male	0	22.53	19.00	0

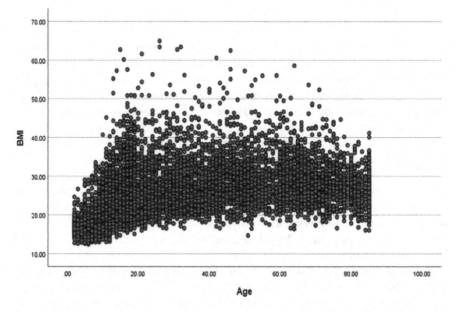

FIGURE 5.1
BMI versus Age

inches), and obesity (BMI above 29.9), among other potentially explanatory factors for each participant.

A graphical display of BMI versus waist size provides a pattern of sloping upwards, as seen in Figure 5.1. This graphical display suggests that there is a positive relationship between BMI and waist size. However, this observation is based on a graphical representation and does not provide statistical evidence. One needs the use of a statistical test to declare whether meaningful differences exist or to declare statistical significance.

5.2 Continuous Response with One Continuous Covariate: Simple Linear Regression

5.2.1 Correlation Coefficient (a Linear Relation)

Let us begin with a simple measure of association (correlation coefficient) between two continuous variables, e.g., BMI and age measurements. The association among variables is closely related to simple linear regression. Linear regression addresses how one covariate (explanatory or independent or X variable), waist size, is associated with the variable of interest (response or outcome or dependent or Y variable), BMI. Association does not specify one variable as response and the other variable as covariate but rather addresses relationships and is measured by a *correlation coefficient*.

Correlation Coefficient is a single value that provides a measure of the strength and direction of the association between two variables. It takes on values between −1.00 and +1.00. A value of −1.00 says that a continuous response variable is perfectly negatively correlated with a continuous explanatory variable. So as one variable increases, the other variable decreases. A value of 0 tells us that there is no linear relationship between the two variables. As one variable changes, the other variable remains the same. A value of +1 denotes that the two continuous variables are perfectly positively correlated, so as one variable increases, the other variable increases. However, the correlation among variables from real data is rarely ever perfect. In the analysis of data the idea is to know how much correlation is enough to say that it exists in a statistically significant amount.

5.2.2 Fundamentals of Linear Regression

A linear regression model provides an approach to identify the impact of an explanatory variable X_1 or a set of explanatory variables $[X_1, X_2, ...X_p]$ on the mean of the response variable of interest Y. If there is one covariate under consideration on one continuous response, it is referred to as simple linear regression. If there is more than one continuous covariate, this is referred to as multiple linear regression.

Each distinct value of the explanatory variable $[X_i]$ represents a subpopulation. Thus, for each value of X, one can find a Y value, $(Y_i | X_i = x_i)$. The set of y_i values for a given x_i value, are seen in Figure 5.2. The response y_i for each x_i value is assumed to have value y_i taken from a normally distributed subpopulation. Although there may be only one y value for each x value in a sample set of data, there is a population of Y_i for each of the X_i values. Those Y_i's are normally distributed as depicted in Figure 5.2 with the assumption that the variance of the Y_i's at each of the x values is the same. Thus, the number of distinct x_i values is equivalent to the number of subpopulations under consideration in the sample of data. The boundaries of

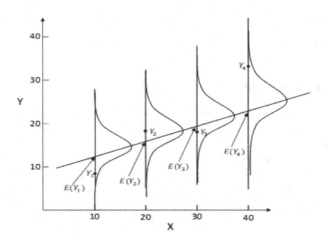

FIGURE 5.2
Y_i versus subpopulations of X_i

these subpopulations [$10 \leq x_i \leq 40$] also provide an area in which prediction may be made, as seen in Figure 5.2. Going beyond the scope to predict can be problematic. Predicting beyond the scope of the data is assuming that the behavior among the variables exists as it did within the scope.

Regression models seek to find a relationship between the mean of the response y and the subpopulation. It relates the mean of y's across the characteristics of different subpopulations. One does not expect to see a perfect relation between mean of Y_i and the X_i. One looks for the best approximation through the method of ordinary least squares (OLS). Due to random variation in each of the subpopulations, it is rare or impossible to find a relationship that will be exactly linear. For example, Figure 5.1 is a two-dimensional figure and not quite a straight line. This figure seems to approximate the means across the subpopulations. So instead of a perfect linear relation, there is a perturbance in ε_i, so that

$$\text{Observation}_i = y_i \mid x_i = \beta_0 + \beta_1 x_i + \varepsilon_i$$

where y_i denotes the ith observation output value, β_0 and β_1 denote two regression parameters that represent the intercept and slope, respectively. The slope [β_1] is also known as a regression parameter or coefficient of X_i and ε_i is the error term. The error term represents things pertinent to an outcome but not necessarily measured or noted. The desire is to obtain the best approximation. Thus, our regression model is an approximation of the relationship between Y and X.

As an example, consider *Age* as an explanatory factor of *BMI*. We want to fit a linear relation. In reality, by fit we mean that as *Age* increases by one unit, one expects BMI to increase or decrease by a certain amount. More specifically, one might ask the following questions:

1. Is there a relationship between BMI and Age?
2. How strong is the relationship between BMI and Age?
3. How accurately can one predict BMI based on Age?
4. Is the relationship linear?

Because of random variation, there will not be a perfect linear relation, that is, a perfect line. However, one may ask for the best straight line based on the sample of data, for $Y_i = \beta_0 + \beta_1 X_i + \varepsilon_i$. One obtains the best approximation through a straight line based on the method of ordinary least-squares. The best line is

$$\hat{y}_i = b_0 + b_1 X_i,$$

Where b_0 and b_1 are OLS estimates based on the sample of data. This is referred to as the OLS equation or the predicted equation. The $\hat{y}_i$s are the model value or the predicted value. Thus, the observed value is (x_i, y_i), and the predicted value is $(x_i, \hat{y}_i)$.

5.2.3 Estimation – Ordinary Least-Squares (OLS)

The method of ordinary least squares (OLS) is used to obtain b_0 and b_1 as estimates of β_0 and β_1. It minimizes the difference between the true value and the predicted value by minimizing the sum of squared error (SSE),

$$\text{SSE} = \sum_i^N \left(Y_i - \beta_0 - \beta_1 X_i\right)^2$$

with respect to β_0 and β_1. The emerging straight line may not go through any of the observed data points, but it will be the closest overall in approximating the observed points. The variation is measured in squared units. It provides the straight line with the best fit. By best, we mean that there is no other line where the sum of squares of the difference between its value and the corresponding observed y_i values is less.

5.2.4 The Coefficient of Determination:

If the y_i values were all the same, one would never question the variation (differences) in Y_i. An examination of the y values reveals that they are not all the same. To explain the difference, one may hypothesize that a certain X covariate or a group of covariates can explain the variation in Y_i. Each X covariate or set of covariates in a model is a possible way to explain the difference in the y values. A measure of how well the covariate or covariates explain or account for the variation in Y_i is referred to as the coefficient of determination or R^2, which measures the proportion of the total variation among the observed values of Y explained by the linear regression. The coefficient of determination lies in the range [0, 1]. In the simple linear regression, the correlation in the sample of data between X and Y is denoted by r_{xy} and as such that $r_{xy}^2 = R^2$. Also, the square root $\sqrt{R^2} = R$. It measures the correlation between the predicted values and the observed values.

5.2.5 Hypothesis Testing

When one examines the Y_i values, it is evident that they are different. Linear regression is a process that allows one to explain the differences in Y_i and make predictions about outcome "$\hat{y}_i$" based on the knowledge gathered about covariate "X". When considering X, one accounts for some of the variation in Y_i but usually not all of it. One has accounted for $\hat{y} = b_0 + b_1 x$. Thus, there is no need to account for Y_i, but one must still account for $Y_i - \hat{Y}_i$. In other words, one needs to account for Y_i after adjustment for X.

The results from the analysis of data using a simple linear regression model are usually organized or summarized in tabular form, as in Table 5.2. The first row deals with the source of variation explained, while the second row deals with the source of variation left unexplained. Adding more X variables into the model decreases the amount of variation accounted for. The next logical question would be to ask by how much.

TABLE 5.2

Summary of the Variation in Regression

Source	DF	Sum of Squares	Mean Squares	F test	p-Value
Model (Explanation in variation in Y_i)	# X's	Total variation in Y_i explained by the covariates	Average of the total variation explained	F = ratio of average of variation explained to the average of the variation unexplained	p-value
Error (what was left unexplained)	n−# betas	Total variation in Y_i not explained by the covariates	Average of the total variation unexplained = variance in Y_i after covariates explained the difference		
Corrected Total	n−1	Total variation in Y_i	Average of the total variation = variance in Y_i		

5.2.6 F-Test Statistic

Consider modeling the mean of Y_is without covariates. Thus, variation in Y_i from the mean is $\sum(Y_i - \bar{Y})^2 = SST$. Consider X as a covariate, which has accounted for $\hat{Y}_i$ i.e. $\sum(\hat{Y}_i - \bar{Y})^2 = SSR$. There is the variation in Y_i left unmeasured by $\sum(Y_i - \hat{Y}_i)^2 = SSE$ (since the variation due to X is $\hat{Y}_i$) left unexplained and unaccounted for. If the slope is zero or near zero then $\hat{Y}_i$ will differ a great deal from the observed Y_i, and there will be a lot of variation left to be explained. As such, a measure (ratio of explained variation by X as opposed to what was left unexplained) is obtained and referred to as an F-test statistic. The hypotheses are:

$$H_0 : \beta_1 = 0$$

slope between Y and X is zero, i.e., X does not explain any variation in Y_i.

$$H_1 : \beta_1 \neq 0$$

slope between Y and X is not zero, so X does explain some variation in Y_i. Although based on OLS principles, the best model is $\hat{y}_i = b_0 + b_1 X_i$, where more steps are often taken to know if it can be improved upon. Some common diagnostic measures are presented.

5.2.7 Diagnostic Measures (Weisberg 1985)

Cook's distance (D_i) is a measure of the influence of an observation based on the total changes in all other residuals when the observation is used versus when it is not used. Large values (usually greater than 1) indicate substantial influence by the observation as it relates to estimating the regression coefficients.

Hat values or leverages provide information about the impact of the x values on the predicted values. If all cases have equal influence, each would have a value of p / n, where p equals the number of regression coefficients, and n is the sample size. If a case has no influence, its value would be $1/n$, whereas total domination by a single case would result in a value of $(n-1)/n$ Values exceeding $2p / n$ for larger samples, or $3p / n$ for smaller samples ($n \leq 30$), are likely observations to be classified as influential observations (Belsley, Kuh, and Welsch 1980).

An outlier is an observation that has a substantial difference between the observed values and predicted values of the dependent variable (a large residual) or between its independent variable values and those of other observations (Barnett and Lewis 1984).

Standardized residual is the difference between the observed value and the predicted value but measured on a standardized scale. It has a mean of 0 and a standard deviation of 1. The standardized residual provides a means of identifying outliers due to Y by locating values larger than 2.0 or less than −2.

Studentized residual is another form of standardized residual. It is based on a comparison of the calculations with and without that observation.

5.2.8 Analysis of Data Using SAS Program

```
proc corr data=nhgh;
var bmi age;
run;
** Simple Linear Regression;
proc reg data=nhgh ;
model bmi=age/p r clm cli covb stb clb;
output out=outreg1 p=predict r=resid RSTUDENT=RSTUDENT
          cookd=cookd;
run;
ods graphics off;
** Check the residuals for normality;
proc univariate data=outreg1 plot normal;
var RSTUDENT;
histogram/normal;
qqplot/normal(mu=est sigma=est); run;
```

The SAS output contains:

The CORR Procedure

Simple Statistics							
Variable	N	Mean	Std Dev	Sum	Minimum	Maximum	Label
BMI	6795	28.32174	6.95011	192446	13.18000	84.87000	BMI
Age	6795	44.28570	20.59459	300921	12.00000	80.00000	Age

There are 6795 observations. The mean BMI is 28.32, with standard deviation of 6.95. The smallest BMI reported is 13.18, and the largest is 84.87.

Pearson Correlation Coefficients, N = 6795
Prob > |r| under H0: Rho=0

	BMI	Age
BMI	1.00000	0.20907
BMI		<.0001
Age	0.20907	1.00000
Age	<.0001	

The correlation between BMI and age is positive with value +0.209. Correlation lies between −1 and +1. The p-value is <.0001, suggesting that the correlation is statistically significant.

The REG Procedure
Dependent Variable: BMI

Number of observations read	6795
Number of observations used	6795

Analysis of Variance

Source	DF	Sum of Squares	Mean Square	F Value	Pr > F
Model	1	14344	14344	310.49	<.0001
Error	6793	313833	46.19950		
Corrected Total	6794	328178			

Root MSE		6.79702	R-Square	0.0437	
Dependent Mean		28.32174	Adj R-Sq	0.0436	
Coeff Var		23.99931			

The total variation in BMI is 328178. Age explains 14344 of that variation. The mean variation explained is 14344. The model variance or the mean variation left over is 46.1995. It is the $var\left((BMI|Age)\right)$

$H_0 : \beta_1 = 0$. The p-value is <.0001, signifying statistically significant correlation between Age and BMI. (See Table 5.2.) However, only 4.37% (coefficient of determination or R square) of the variation in Y is explained by Age. There is room for other covariates to enter.

Variable	DF	Parameter Estimate	Standard Error	t Value	Pr > |t|	Standardized Estimate	95% Confidence Limits	
Intercept	1	25.197	0.196	128.85	<.0001	0	24.813	25.581
age	1	0.071	0.004	17.62	<.0001	0.209	0.063	0.078

The best straight line that fits the data is $\widehat{BMI} = 25.197 + 0.071Age$. The slope 0.071 has a standard error of 0.004, producing a t value = 0.071/0.004 = 17.62 and p-value <.0001. The 95% confidence limits are a range of possible values for β_1 [0.0627, 0.0784].

Obs	Dependent Variable	Predicted Value	95% CL Mean		Residual	Student Residual	-2-1 0 1 2	Cook's D	RStudent	Hat Diag H
1	32.2	27.608	27.428	27.788	4.612	0.679	| |* |	0.000	0.679	0.0002
2	22.0	26.385	26.116	26.654	-4.385	-0.645	| *| |	0.000	-0.645	0.0004
3	42.4	29.442	29.238	29.646	12.948	1.905	| |*** |	0.000	1.906	0.0002
4	32.6	27.038	26.822	27.253	5.573	0.820	| |* |	0.000	0.820	0.0003
5	30.6	28.701	28.534	28.869	1.869	0.275	| | |	0.000	0.275	0.0002
6	26.0	30.842	30.518	31.165	-4.802	-0.707	| *| |	0.000	-0.707	0.0006
7	27.6	30.842	30.518	31.165	-3.222	-0.474	| | |	0.000	-0.474	0.0006
8	26.0	26.426	26.160	26.692	-0.456	-0.067	| | |	0.000	-0.067	0.0004
9	16.6	26.114	25.820	26.408	-9.514	-1.400	| **| |	0.000	-1.400	0.0005
10	39.9	28.231	28.069	28.393	11.669	1.717	| |*** |	0.000	1.717	0.0001

> The dependent variable or the Y value for observation #1 is 32.2. The predicted value from the fitted model is 27.61, and the confidence interval is [27.43, 27.79] for the mean. The difference between what the model predicted 27.61 and the observed value 32.2 is the residual 4.61. When one standardizes the residual, one gets a value of 0.679. It is within the range of [-2, 2] so it is not statistically significant. Hence, one is satisfied with the model as it pertains to that data point. The leverage (Hat Diag) is 0.0002, so it is not an outlier due to its x values. Its Cook's D is 0.000, so the point is not influential.

5.2.9 Analysis of Data Using R Program

```
## load the R library "readxl" to read Excel data file
library(readxl)
nhgh = read_excel(("C:/Users/angel/Desktop/new_nhgh.
                   xls"))
# Calculate the correlation
cor(nhgh$bmi,nhgh$age)
## [1] 0.2090676
## Simple linear regression and LSE;
model1 =lm(nhgh$bmi~nhgh$age)
summary(model1)
## Call:
## lm(formula = nhgh$bmi ~ nhgh$age)
## Residuals:
##    Min    1Q    Median 3Q      Max
## -17.362  -4.616  -1.198 3.407  56.416
## Coefficients:
##             Estimate  Std. Error t value Pr(>|t|)
## (Intercept) 25.197182 0.195557   128.85  <2e-16   ***
```

```
## nhgh$age     0.070555   0.004004   17.62   <2e-16   ***
## Signif. codes: 0 '***' 0.001 '**' 0.01 '*' 0.05 '.'
0.1 ' ' 1
```

> The best straight line fitted to the data is $\widehat{BMI} = 25.197 + 0.071 Age$.
> The slope 0.071 has a standard error of 0.004, producing a t value =
> $0.071/0.004 = 17.62$. The p-value is <2e-16 ***.

```
## Residual standard error: 6.797 on 6793 degrees of
                 freedom
## Multiple R-squared: 0.04371,   Adjusted
                 R-squared: 0.04357
## F-statistic: 310.5 on 1 and 6793 DF, p-value: <
                 2.2e-16
```

> $H_0 : \beta_1 = 0$. The p-value <2e-16, signifying statistically significant cor-
> relation between Age and BMI (see Table 5.2). However, only 4.37%
> (coefficient of determination) of the variation in Y is explained by Age.
> There is room for other covariates to be included.

```
##     Dep     Phat     Lwr       Upr leverage  Resid StudResid  CooksD
## 1   32.22  27.60780  27.42770  27.78790  0.00018   4.61220   0.67860   0.00004
## 2   22.00  26.38485  26.11548  26.65422  0.00041  -4.38485  -0.64522   0.00009
## 3   42.39  29.44222  29.23809  29.64634  0.00023  12.94778   1.90551   0.00043
## 4   32.61  27.03748  26.82175  27.25321  0.00026   5.57252   0.81993   0.00009
## 5   30.57  28.70139  28.53433  28.86846  0.00016   1.86861   0.27492   0.00001
## 6   26.04  30.84155  30.51796  31.16514  0.00059  -4.80155  -0.70660   0.00015
## 7   27.62  30.84155  30.51796  31.16514  0.00059  -3.22155  -0.47408   0.00007
## 8   25.97  26.42601  26.16029  26.69173  0.00040  -0.45601  -0.06710   0.00000
## 9   16.60  26.11439  25.82040  26.40838  0.00049  -9.51439  -1.40023   0.00048
## 10  39.90  28.23103  28.06907  28.39298  0.00015  11.66897   1.71715   0.00022
```

> The dependent variable or the Y value for observation #1 is 32.2. The
> predicted value from the fitted model is 27.61 and the confidence inter-
> val is [27.43, 27.79] for the mean. The difference between what the model
> predicted 27.61 and the observed value 32.2 is the residual 4.61. When
> one standardizes the residual, one gets a value of 0.679. It is within the
> range of [−2, 2] so it is not statistically significant. Hence, one is satisfied
> with the model fit as it pertains to that data point. The leverage (Hat
> Diag) is 0.0002, so it is not an outlier due to its x values. Its Cook's D is
> 0.000, so the point is not influential.

Comment on assessing the accuracy of the coefficient estimates: The standard error of an estimator reflects the fluctuation about the mean under-repeated sampling. However, in practice, one never really observes an under-repeated sampling, but estimates it based on one sample of data. The standard errors are important for computing confidence intervals. A 95% confidence interval is defined as a range of possible values to which one attaches 95% degree of belief that it includes the true value of the parameter. However, this interval either includes or does not include the value. There is no probability attached to that one event. Standard errors are also used to perform hypothesis tests on the coefficients of the model.

5.3 Continuous Responses with Multiple Factors: Multiple Linear Regression

Multiple regression models allow one to examine the simultaneous effect of multiple factors on a continuous response as shown in Figure 5.3, where two predictors (i.e., "Age" and size of "Waist") are used to model the mean outcome of BMI. The set of factors (multiple covariates) may account for some of the variation in the response outcomes.

There are several predictors competing to provide an explanation for some of the variation in Y. At times, the predictors are correlated, so one must address the issue of multicollinearity or significant correlations among predictors. Multicollinearity is not an assumption. It is a desire that only matters if one wants to interpret the contribution of a covariate within a system of covariates. Multicollinearity is not a problem if one is interested in prediction Draper and Smith (1998).

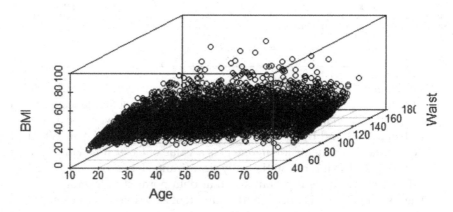

FIGURE 5.3
Illustration of multiple regression of BMI with age and waist

The model in unit form as each observation within a subpopulation is

$$Y_i = \beta_0 + \beta_1 X_{1i} + \beta_2 X_{2i} + \beta_3 X_{3i} + \varepsilon_i.$$

where μ_i represents the random variation. In particular, in terms of one observation is

$$BMI_i = \beta_0 + \beta_1 Age_i + \beta_2 Waist_i + \beta_3 Albumin_i + \varepsilon_i.$$

While in the form of mean, our model is written as the mean of Y $\mu_{Y|X}$ for each subpopulation X as follows:

$$\mu_{Y|X} = \beta_0 + \beta_1 X_{1i} + \beta_2 X_{2i} + \beta_3 X_{3i}$$

The mean model has no μ_i term as the mean of μ_i is assumed to be zero.

Interpreting regression coefficients: Interpret β_j as the average effect on Y of one unit increase in X_j, holding all other predictors fixed. By fixed we mean that the values of the other predictors remain the same while the one predictor changes. In the BMI example, the model

$$BMI_i = \beta_0 + \beta_1 Age_i + \beta_2 Waist_i + \beta_3 Albumin_i + \varepsilon_i$$

is in a four-dimensional setting as there are three regression variables and one response. The least-squares regression plane is in four-dimensions, so it is a matter of finding the best plane. The plane is chosen to minimize the sum of the squared distances between each observation and its predicted value (from the model) in the plane.

The ideal scenario in a multiple linear regression model is when the predictors in the model are completely uncorrelated, but that is seldom the case. If that were the case, then each coefficient can be estimated and tested separately, and their coefficient would be the same as in the simple linear regression (by itself). More so, the interpretations such as "a unit change in X_j is associated with a β_j change in Y_i, while all the other variables are held fixed", would be the reality. While a little correlation is not a problem, significant correlations among the predictors (multicollinearity) can cause serious problems. The problem is that variance of estimators is large. This results in wide confidence intervals and large p-values. Hence, one is likely to say a covariate is not statistically significant in the model when, in fact, it may be. The interpretation of a coefficient under such circumstances is difficult to comprehend holding a covariate fixed as when X_j changes and everything else stays fixed. Two measures for checking significant multicollinearity are **tolerance** and **variance inflation factor**.

Tolerance is used to measure multicollinearity. Tolerance values approaching zero indicate that the variable is highly correlated with one or more of the other predictor variables.

Variance inflation factor (VIF) is a measure of the effect of other predictor variables on a regression coefficient. VIF is inversely related to the tolerance value. The VIF reflects the extent to which the standard error of the regression coefficient is increased due to multicollinearity. Large VIF values greater than 10.0, (which corresponds to a tolerance of .10) indicate a high degree of collinearity or multicollinearity among the independent variables.

5.3.1 Analysis of Data Using SAS Program

```
** MULTIPLE LINEAR REGRESSION;
** INPUT AGE WAIST ALBUMIN & OUTPUT BMI;
** CONTINUOUS;
proc reg data=nhgh;
model bmi=age waist albumin/p r clm cli covb stb clb;
output out=outreg2 p=predict r=resid RSTUDENT=RSTUDENT
          cookd=cookd;
run;
proc univariate data=outreg2 plot normal;
var RSTUDENT;
histogram/normal;
qqplot/normal(mu=est sigma=est);
run;
```

The REG Procedure

Model: MODEL1
Dependent Variable: BMI

Number of **observations read**	6795
Number of **observations used**	6475
Number of **observations with missing values**	320

Analysis of Variance					
Source	DF	Sum of Squares	Mean Square	F Value	Pr > F
Model	3	253897	84632	13203.9	<.0001
Error	6471	41477	6.40963		
Corrected Total	6474	295374			
Root MSE	2.532		R-Square		0.8596
Dependent Mean	28.1886		Adj R-Sq		0.8595
Coeff Var	8.981				

There are 6475 observations. The total variation in BMI is 295374. Age, waist, and albumin account for 253897 of the total variation. There is still 41477 to be explained. The mean variation explained is 84632. The mean variation left to be explained is 6.41. The model variance is 6.41. The standard error is $2.532 = \sqrt{(6.41)}$. The coefficient of determination is 85.95%.

$H_0 : \beta_1 = \beta_2 = \beta_3 = 0$. The p-value is <.0001, signifying statistically significant. There is correlation between Age, waist size, and albumin and BMI (see Table 5.2). Together they explain 85.96% (coefficient of determination) of the variation in Y. The correlation between the observed and the predicted is $R = \sqrt{0.8596} = 0.9271$.

Parameter Estimates

Variable	Label	DF	Parameter Estimate	Standard Error	t Value	Pr > |t|	Standardized Estimate	95% Confidence Limits	
Intercept	Intercept	1	2.157	0.541	3.98	<.0001	0	1.096	3.219
Age	Age	1	-0.054	0.002	-32.21	<.0001	-0.165	-0.06	-0.05
Waist	Waist	1	0.376	0.002	182.76	<.0001	0.947	0.372	0.379
Albumin	Albumin	1	-1.804	0.105	-17.24	<.0001	-0.08644	-2.01	-1.60

The fitted equation is $\widehat{BMI} = 25.157 - 0.054 Age + 0.376 Waist -1.804 Albumin$. Each of the variables in the model is statistically significant. This is seen through the p-value of <.0001. The 95% confidence limits do not include zero, as seen in [−0.06, −0.05] for Age, [0.372, 0.379] for Waist, and [−2.01, −1.60] for Albumin.

5.3.2 Analysis of Data Using R Program

```
## Call "lm" to fit the multiple linear regression
model2 = lm(nhgh$bmi~nhgh$age+nhgh$waist+nhgh$albumin)
summary(model2)
## Call:
## lm(formula = nhgh$bmi ~ nhgh$age + nhgh$waist + nhgh$albumin)
## Residuals:
##     Min        1Q      Median     3Q      Max
##    -11.602    -1.613    -0.161    1.369   35.145
## Coefficients:
##               Estimate     Std. Error   t value    Pr(>|t|)
```

```
## (Intercept)     2.157392     0.541418     3.985     6.83e-05  ***
## nhgh$age        -0.054258     0.001684    -32.210   < 2e-16   ***
## nhgh$waist       0.375664     0.002056    182.758   < 2e-16   ***
## nhgh$albumin    -1.804125     0.104676    -17.235   < 2e-16   ***
## Signif. codes: 0 '***' 0.001 '**' 0.01 '*' 0.05 '.' 0.1 ' ' 1
```

The fitted equation is

$$\widehat{BMI} = 25.157 - 0.054Age + 0.376Waist - 1.804Albumin.$$

Each variable in the model is statistically significant. This is seen through the p-value of <.0001.

```
## Residual standard error: 2.532 on 6471 degrees of
##            freedom
##  (320 observations deleted due to missingness)
## Multiple R-squared: 0.8596, Adjusted R-squared: 0.8595
## F-statistic: 1.32e+04 on 3 and 6471 DF, p-value: <
##            2.2e-16
plot(model2)
```

$H_0 : \beta_1 = \beta_2 = \beta_3 = 0$. The p-value is <.0001, signifying statistically significant simultaneous correlation between Age, Waist, and Albumin and BMI (see Table 5.2.) Together they explain 85.95% (coefficient of determination) of the variation in Y. The correlation between the observed and the predicted is R = $\sqrt{0.8596}$ = 0.9271.

The fitted equation is $\widehat{BMI} = 25.157 - 0.054Age + 0.376Waist$ $-1.804Albumin$. Each variable in the model is statistically significant. This is seen through the p-value of <.0001. The 95% confidence limits do not include zero as seen [−0.0576, −0.051] for Age, [0.372, 0.379] for Waist size, and [−2.01, −1.60] for Albumin.

5.4 Continuous Responses with Continuous and Qualitative Predictors

Consider modeling the mean of the normal random variable, but instead of using quantitative predictors only, we use a combination of quantitative and qualitative predictors. When predictors are qualitative, taking a discrete set of distinct values, they are referred to as categorical predictors or factors. When the predictor has only two levels, we refer to it as a binary predictor (for example, gender). For a binary predictor, one can use a dummy coding of zero-one variable.

5.4.1 Qualitative Predictors with More Than Two Levels

One approach to add qualitative predictors with more than two levels to the model, is to create a set of dummy variables, one for each category. For example, for the ethnicity variable with five categories, we create four dummy variables for $5 - 1 = 4$. Let $D_1 = 1$ if the ith person is Mexican, $D_2 = 1$ if Hispanic, $D_3 = 1$ if Caucasian, $D_4 = 1$ if Black, $D_5 = 1$ and 0 otherwise. The following data structure would be coded as:

ID	Race	D_1	D_2	D_3	D_4	D_5
001	Mexican	1	0	0	0	0
002	Hispanic	0	1	0	0	0
003	Caucasian	0	0	1	0	0
004	Black	0	0	0	1	0
005	Other	0	0	0	0	1

However, this kind of structure presents a non-estimable situation. This is because $1 = D_1 + D_2 + D_3 + D_4 + D_5$. To avoid this and have estimable coefficients, we can omit one of the D_i. The omitted D_i is realized as it is identified when all the other D_i are zero. The model is

$$\mu_{Y|X} = \beta_0 + \beta_1 X_{1i} + \beta_2 X_{2i} + \beta_3 X_{3i} + \gamma_1 D_1 + \gamma_2 D_2 + \gamma_3 D_3 + \gamma_4 D_4.$$

The $D_5 = 1$ if i^{th} person is not Mexican, Hispanic, Caucasian, or Black. Thus, the model provides an estimate of D_5 without having it in the model. It is represented when $D_1, D_2, D_3,$ or D_4 are zero. Thus, when all the dummy variables are zero, then the "other" is the baseline. It is convenient when setting up the model to choose the best category to be baseline based on the needed conclusions or interpretations (Rousseeuw and Leroy 1987).

5.4.2 Analysis of Data Using SAS Program

The SAS program is as follows:

```
data new_nhgh;
set nhgh;
gender2=(gender='Male');
```

```
Mexican=(re='Mexican American');
Hispanicn=(re='Other Hispanic');
White=(re='Non-Hispanic White');
Black=(re='Non-Hispanic Black');
run;
proc reg data=new_nhgh;
model bmi=age waist albumin Mexican Hispanicn White
          Black/p r clm cli covb stb clb;
run;
```

The SAS output contains:

The REG Procedure
Dependent Variable: BMI

Number of observations read	6795
Number of observations used	6475
Number of observations with missing values	320

Analysis of Variance

Source	DF	Sum of Squares	Mean Square	F Value	Pr > F
Model	7	255253	36465	5877.63	<.0001
Error	6467	40121	6.20397		
Corrected Total	6474	295374			
Root MSE		2.49078	R-Square	0.8642	
Dependent Mean		28.18846	Adj R-Sq	0.8640	
Coeff Var		8.83616			

There are 6475 observations. The total variation in BMI is 295374. The 7 covariates account for 255253 of the total variation. There is still 40121 to be explained. The mean variation explained is 36465. The mean variation left to be explained is 6.41. The model variance is 6.204. The standard error is $2.49 = \sqrt{(6.20)}$. The coefficient of determination is 86.42%.

$H_0 : \beta_{age} = \beta_{waist} = \beta_{albumin} = \beta_{Mexican} = \beta_{Hispanic} = \beta_{White} = \beta_{Blacks} = 0$ versus they are not all equal to zero. The p-value <.0001, signifying statistically significant. The correlation between Age, Waist, Albumin, and Ethnicity and BMI. (See Table 5.2.) Together they explain 86.42% (coefficient of determination) of the variation in Y. The correlation between the observed and the predicted is $R = \sqrt{0.8642} = 0.927$.

Variable	Label	Parameter Estimate	Standard Error	t Value	Pr > |t|	Standardized Estimate	95% Confidence Limits	
Intercept	Intercept	0.84322	0.55184	1.53	0.1266	0	-0.23856	1.92501
Age	Age	-0.05015	0.00169	-29.71	<.0001	-0.15220	-0.05346	-0.04684
Waist	Waist	0.37619	0.00203	185.16	<.0001	0.94803	0.37221	0.38017
Albumin	Albumin	-1.55412	0.10444	-14.88	<.0001	-0.07446	-1.75885	-1.34940
D_1 =Mexican	Mexican	0.19528	0.14759	1.32	0.1858	0.01167	-0.09405	0.48462
D_2 =Hispanic	Hispanic	0.21097	0.16179	1.30	0.1923	0.00956	-0.10620	0.52814
D_3=White	White	-0.42493	0.13888	-3.06	0.0022	-0.03136	-0.69719	-0.15267
D_4 =Black	Black	0.84447	0.15084	5.60	<.0001	0.04737	0.54878	1.14015

$H_0 : \beta_{age|\text{waist albumin Mexican Hispanic, Whites and blacks}} = 0$ versus it is not equal to zero. The p-value is <.0001, indicating statistically significant correlation between BMI and Age, after adjusting for waist, albumin and Ethnicity The 95% confidence limits do not include zero [−0.053, −0.047]. (See Table 5.2.) The equation is

$\widehat{BMI} = 0.843 - 0.050Age + 0.376Waist - 1.554Albumin + 0.195Mexican + 0.211Hispanic - 0.425White + 0.844Black.$

The variables Age, Waist, and Albumin in the model are statistically significant. The ethnicity group showed White and Black differ significantly in their BMI.

5.4.3 Analysis of Data Using R Program

```
model3 = lm(bmi~age+waist+albumin+Mexican+Hispanic+White+Black,
data=nhgh)
summary(model3)
##
## Call:
## lm(formula=bmi~age+waist+albumin+Mexican+Hispanic+White+Black,
            data=nhgh)
##
## Residuals:
##   Min     1Q    Median  3Q     Max
## -11.293 -1.558 -0.165  1.311  34.345
##
## Coefficients:
##              Estimate   Std. Error  t value   Pr(>|t|)
## (Intercept) 0.843225   0.551838    1.528     0.12655
## age         -0.050153  0.001688    -29.713   < 2e-16 ***
## waist        0.376191  0.002032    185.156   < 2e-16 ***
## albumin     -1.554125  0.104435    -14.881   < 2e-16 ***
## Mexican      0.195285  0.147593    1.323     0.18584
## Hispanic     0.210967  0.161793    1.304     0.19230
```

```
## White      -0.424932   0.138884    -3.060    0.00223 **
## Black       0.844467   0.150836     5.599    2.25e-08 ***
```

$H_0 : \beta_{age|\text{waist albumin Mexican Hispanic, Whites and blacks}} = 0$ versus it is not equal to zero. The p-value $<.0001$, indicating statistically significant between BMI and Age, after adjusting for Waist, Albumin and Ethnicity. The 95% CI does not include zero $[-0.053, -0.047]$.

$\widehat{BMI} = 0.843 - 0.050 Age + 0.376 Waist - 1.554 Albumin + 0.195 Mexican + 0.211 Hispanic - 0.425 White + 0.844 Black.$

The variables Age, Waist, and Albumin in the model are statistically significant. The ethnicity group showed White and Black differ statistically significantly in their BMI.

```
## Residual standard error: 2.491 on 6467 degrees of
##            freedom
##     (320 observations deleted due to missingness)
## Multiple R-squared: 0.8642, Adjusted R-squared: 0.864
## F-statistic: 5878 on 7 and 6467 DF, p-value: < 2.2e-16
```

$H_0 : \beta_{age} = \beta_{waist} = \beta_{albumin} = \beta_{Mexican} = \beta_{Hispanic} = \beta_{White} = \beta_{Blacks} = 0$ versus they are not all equal to zero. The p-value $<.0001$, indicating statistically significant simultaneous correlation between BMI and Age, Waist, Albumin, and Race. (See Table 5.2.) Together they explained 86.42% (coefficient of determination) of the variation in Y. The correlation between the observed and the predicted is $R = \sqrt{0.8642} = 0.927$.

5.5 Continuous Responses with Continuous Predictors: Stepwise Regression

The ultimate goal of a regression model is the development of a regression equation (line or plane of best fit) between the response and several covariates. The research question may lead to a test of predictor variables to be considered for entry into the regression model. A stepwise regression helps provide the best combination of predictors to model the outcome. Predictors are entered into the regression equation one at a time based upon some statistical criterion. At each step of the analysis, the predictor variable that contributes the most to the prediction equation in terms of increasing the multiple correlation, R, is entered. This process is continued only if additional variables add statistically to the regression equation. When no additional predictor variable adds anything statistically meaningful to the regression equation, the analysis stops. Thus, not all covariates may enter the equation in stepwise regression.

5.5.1 Analysis of Data Using SAS Program

```
*** Stepwise;
PROC REG data=nhgh;
MODEL bmi=waist age albumin SCr bun/SELECTION=FORWARD
      SLSTAY=0.3;
run;
```

> In the program, one can specify a criterion for a variable that came into the model to stay in the model, such as [SLSTAY=0.3;].

The SAS output contains:

Dependent Variable: BMI

Number of observations read	6795
Number of observations used	6475
Number of observations with missing values	320

Forward Selection: Step 1

Variable waist Entered: R-Square = 0.8342 and C(p) = 1220.789

Analysis of Variance

Source	DF	Sum of Squares	Mean Square	F Value	Pr > F
Model	1	246415	246415	32579.4	<.0001
Error	6473	48959	7.56353		
Corrected Total	6474	295374			

Variable	Parameter Estimate	Standard Error	Type II SS	F Value	Pr > F
Intercept	−6.63986	0.18465	8666.74651	1293.00	<.0001
Waist	0.38437	0.00204	238543	35588.4	<.0001
Age	−0.04881	0.00169	5577.95554	832.18	<.0001

Forward Selection: Step 2					
Variable age Entered: R-Square = 0.8531 and C(p) = 346.4494					
Analysis of Variance					
Source	DF	Sum of Squares	Mean Square	F Value	Pr > F
Model	2	251993	125996	18797.5	<.0001
Error	6472	43381	6.70284		
Corrected Total	6474	295374			

At each stage, there is a check to see if the next significant variable should be added to the model.

Variable	Parameter Estimate	Standard Error	Type II SS	F Value	Pr > F
Intercept	−6.63986	0.18465	8666.74651	1293.00	<.0001
waist	0.38437	0.00204	238543	35588.4	<.0001
age	−0.04881	0.00169	5577.95554	832.18	<.0001

Summary of Forward Selection								
Step	Variable Entered	Label	Number Vars In	Partial R-Square	ModelR-Square	C(p)	F Value	Pr > F
1	waist	waist	1	0.8342	0.8342	1220.79	32579.4	<.0001
2	age	age	2	0.0189	0.8531	346.449	832.18	<.0001
3	albumin	albumin	3	0.0064	0.8596	49.3124	297.06	<.0001
4	SCr	SCr	4	0.0009	0.8605	9.3102	41.97	<.0001
5	bun	bun	5	0.0001	0.8606	6.0000	5.31	0.0212

5.5.2 Analysis of Data Using R Program

```
library(MASS)
fit = lm(bmi~waist+age+albumin+SCr+bun,data=nhgh)
s = stepAIC(fit,direction="backward")
## Start:   AIC=11990.08
## bmi ~ waist + age + albumin + SCr + bun
##
##              Df Sum of Sq  RSS      AIC
```

```
## <none>                    41176  11990
## - bun      1      34      41209  11993
## - SCr      1      295     41471  12034
## - albumin  1      1875    43050  12276
## - age      1      5398    46574  12786
## - waist    1      214378  255553 23809
```

> At each stage, there is a check to see if the next least significant variable can stay in the model. It continues until there is no other significant variable to bring into the model.

5.6 Research Questions and Comments

While tools, such as stepwise regression, are used to determine the best model, one must not ignore the theory or research within the discipline to help determine the key predictors. One cannot rely on statistics alone when building a model. For example, if you were modeling blood pressure using some observational data and found that gender was not significant, should you take the binary predictor (gender) out of your model? Probably not, since research tells us that gender has differential effects on blood pressure.

5.7 Exercises

Using the National Health and Nutrition Examination Survey data, do the following:

1. Using BMI as the response variable, perform *stepwise regression* to fit the best model from the following data: age, income, weight, height, waist, leg, and albumin.
2. Using the standardized betas, determine the most significant predictor.
3. Is there multicollinearity in the best model obtained from stepwise regression?
4. How well does the model fit the data?
5. Summarize your findings so an educated person who knows no statistics can understand.

References

Barnett, V., Lewis, T.: *Outliers in Statistical Data* (2d ed.). Wiley, New York (1984)

Belsley, D.A., Kuh, E., Welsch, R.E.: *Regression Diagnostics: Identifying Influential Data and Sources of Collinearity*. Wiley, New York (1980)

Draper, N.R., Smith, H.: *Applied Regression Analysis* (3rd ed.). Wiley, New York (1998)

Rousseeuw, P.J., Leroy, A.M.: *Robust Regression and Outlier Detection*. Wiley, New York (1987)

Weisberg, S.: *Applied Linear Regression*. Wiley, New York (1985)

6

Modeling Continuous Responses with Categorical and Continuous Covariates: One-Way Analysis of Covariance (ANCOVA)

6.1 Research Interest/Question

This chapter examines the impact of a factor (categorical measure) on the mean of normal responses, after adjusting for a covariate (a continuous measure). One may be interested in a research question, such as, "In a clinical study, does a certain type of drug (categorical) have an impact on the HDL level (continuous)?" However, one may suspect or think that the age (covariate) of the patient makes a difference in the HDL level. Alternatively, one may wish to know if there is an interaction between age (covariate) and type of drug (categorical). Does the effect of the drug change as age increases? These kinds of questions are answered with the use of an analysis of covariance model (ANCOVA).

6.2 Public Health Right Heart Catheterization Data

This chapter uses information from the Right Heart Catheterization (RHC) dataset (Connors et al. 1996) to demonstrate the use of ANCOVA models. The effectiveness of RHC in the initial care of critically ill patients is modeled. The dataset consists of information obtained on the first day of hospitalization. A subset of the data is shown in Table 6.1. The response or variable of interest is the mean blood pressure (i.e., "MEANBP1" in Table 6.1). The other variables in Table 6.1 are death within 180 days, age, sex, creatinine, and stage of cancer. In particular, the interest is in the variation in blood pressure across three cancer groups (i.e., "CA" with categories of "no", "yes", and "metastatic").

DOI: 10.1201/9781003315674-6

TABLE 6.1

Subset of Right Heart Catheterization Data

Patient	ca (Cancer)	death (Death up to 180 Days)	age	sex	meanbp1 (blood pressure)	crea1 (Creatinine)
1	Yes	No	70.251	Male	41	1.2
2	No	Yes	78.179	Female	63	0.60
3	Yes	No	46.092	Female	57	2.60
4	No	Yes	75.332	Female	55	1.7
5	No	Yes	67.91	Male	65	3.60
6	No	No	86.078	Female	115	1.40
7	Metastatic	No	54.968	Male	67	1
8	No	Yes	43.639	Male	128	0.7
9	Yes	No	18.042	Female	53	1.7
10	Yes	No	48.424	Female	73	0.5
11	No	No	34.442	Male	66	0.5
12	No	No	68.348	Male	50	0.4
13	No	Yes	74.71	Male	53	4.90
14	No	Yes	42.237	Female	77	1.90
15	No	No	81.971	Male	67	3

TABLE 6.2

Summary Statistics of Age by Cancer Group

Cancer	Mean	Std. Deviation	N=#patients
Metastatic	70.055	35.752	384
No	80.363	38.617	4379
Yes	73.561	35.379	972
Total	78.520	38.047	5735

The summary statistics for the mean blood pressure by cancer categories are presented in Table 6.2. As seen in Table 6.2, the mean blood pressure is highest in the cancer group category "no", with a mean of 80.363, and lowest in the cancer group category "metastatic", with a mean of 70.055.

Figure 6.1 displays blood pressure by creatinine (Crea1) for each level of cancer (metastatic, no, yes). It appears that the lines relating the blood pressure to creatinine are not parallel for each of the groups (metastatic, no, yes). When there is no interaction present between creatinine and cancer groups, the lines should be or near to parallel. The interest is in blood pressure as it differs based on cancer groups, but creatinine is an important covariate although treated as nuisance. A one-way ANCOVA model serves as a test of the main effects of the factor (i.e., cancer) while controlling for the effects of the covariate (i.e., creatinine).

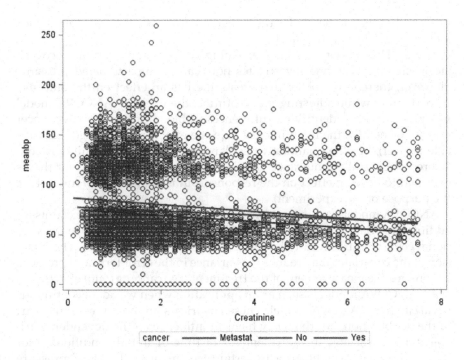

FIGURE 6.1
Graph of blood pressure versus creatinine across cancer

6.3 Hypothesis

For the research question "Does blood pressure have a differential effect across cancer groups?", there are two hypotheses: the null hypothesis

$$H_0 : \mu_{CA1} = \mu_{CA2} = \mu_{CA3}$$

and the alternative or research hypothesis

$$H_1 : \text{not all equal}$$

In this section, a statistical test is outlined for testing the null hypothesis of equal means. It is based on a model consisting of a continuous response variable with a grouping variable (cancer group) and a covariate (creatinine). The main interest is in whether mean blood pressures differ among the cancer groups (Muller and Fetterman 2002).

6.3.1 One-Way ANCOVA Model

The standard one-way ANCOVA model consists of directly incorporating a continuous covariate into an ANOVA model. It improves the power (the probability that one concludes the research hypothesis is true when, in fact, it is true) to detect group effects.

In the RHC dataset, there is a grouping variable in cancer and a covariate in creatinine as one investigates how cancer impacts blood pressure. However, there is the belief that creatinine has an effect on the outcome. In particular, while adjusting for creatinine, the one-way ANCOVA model is a procedure for identifying differences in blood pressure across cancer groups. Generally, the main interest is in the effects of the categorical variable (cancer), but the quantitative explanatory variable (creatinine) is considered a "control" or nuisance variable. (Nuisance factor is a factor that is expected to have an effect on the response, but is not a factor of interest for the purpose of the experiment).

ANCOVA model is an improvement over ANOVA model, as it accounts for additional variation. It augments the ANOVA model with one or more additional covariates believed to be related to the response variable. The inclusion of the covariates may reduce the variance in the error term and provides a more precise measurement of the treatment (i.e., cancer group) effects.

An ANCOVA model is seen as an application when we add a continuous covariate to an ANOVA model. These covariates are not an essential part of the deliberation, but they may have an influence on the dependent variable. Such covariates include elements of time or initial conditions prior to an event (e.g., weight prior to certain treatment or the blood pressure prior to being placed on losartan). Thus the ANCOVA model can be seen as the combination of a one-way analysis of variance model and a regression model:

$$\text{ANCOVA} = \{\text{One way ANOVA}\} + \{\text{Regression}\}$$

6.3.2 Assumptions for ANCOVA

Similar to the one-way ANOVA model, the one-way ANCOVA model allows us to determine whether there are any significant differences between two or more independent (unrelated) groups as they relate to the response. The major difference is the ANOVA model deciphers differences in the group means of response, while the ANCOVA deciphers differences in the adjusted means (i.e., adjusted for the covariate) of response. The one-way ANCOVA model has the additional benefit of allowing one to "statistically control" for a covariate (often referred to as a "nuisance variable"), which one believes will affect the results. This nuisance variable is referred to as the covariate in the model. An ANCOVA model can have more than one covariate. When using any ANCOVA model, it is essential that the assumptions are satisfied or nearly satisfied (Kirk 1968). As is the case for any statistical test, there are several key assumptions that influence the use of an ANCOVA model and leads to the interpretation of the results. The assumptions used in linear regression models and the assumptions in ANOVA models also hold for ANCOVA models.

In particular, one assumes:

- That the response of interest is measured on a continuum;
- Interest is in the differences in the group means of response;
- The observations of the response variable are independent;
- The response is normally distributed, although researchers have reported that approximately normal is enough because it is quite "robust" to violations of normality (robust means that the assumption can be violated to a degree and still provide valid results);
- Variation in response is the same across groups (homoscedasticity);
- The covariate is linearly related to the response in all groups;
- There is no interaction between the covariate and the response variable. This means that the linear regressions of response within each group are parallel (have the same slope).

6.3.2.1 What If We Do Not Have All These assumptions Satisfied?

If the normal assumption does not hold for the response variable, then one can (among other things):

1. Use a transformation of the response variable (e.g., blood pressure);
2. Fit a generalized linear model (a common approach discussed in Chapter 8, where the response has a distribution from the exponential family, the observations are independent, and variance is related to the mean);
3. Use a nonparametric test (no distributional assumption is made about the responses, but the responses are on a scale that one can rank).

6.3.3 F-Test Statistic in ANCOVA Model

In ANCOVA, the covariate is treated as it is in a regression equation. The one-way ANCOVA model is represented as

$$Y_{ij} = \mu + \alpha_i + \beta\left(X_{ij} - \overline{X}_{..}\right) + \varepsilon_{ij}$$

where y_{ij} represents the jth observation in the ith group, α_i is the ith group effect, ε_{ij} denotes the error for the jth observation in the ith group in observing Y_{ij}, β is the parameter of the linear relationship of the covariate X_{ij} on Y_{ij}, and $\overline{X}_{..}$ is the overall mean of the covariate X_{ij}. The factors α_i and the X_{ij} are not related (Doncaster and Davey 2007). Assuming that ε_{ij} are distributed normally with mean zero and variance σ^2, means that the y_{ij} are normally distributed with mean μ_i and constant variance σ^2. The model may also be given as

$$Y_{ij} = \mu_i + \beta\left(X_{ij} - \bar{X}_{..}\right) + \varepsilon_{ij},$$

where the group mean μ_i is reparameterized as an overall mean μ and group effect α_i, such that

$$\alpha_i \rightarrow \mu_i = \mu + \alpha_i.$$

To provide a statistical test for the null hypothesis that the adjusted group means are equal and address the research question, there is a need to construct a measure that takes the key information (group mean, overall mean, sample size, standard error, etc.) into account from the sample of patients while measuring the overall mean difference on a certain scale. This measure is provided through a test statistic value, known as an F-test statistic.

6.3.4 ANCOVA Hypothesis Tests and the Analysis of Variance Table

The hypothesis tests for the ANCOVA model are very similar to an ANOVA model. Their key difference is that the population means for each treatment are adjusted for the covariate. The variance is partitioned as in an ANOVA model but with three components:

1. The treatment variance, reflecting the differing group means, e.g., the different state of cancer;
2. The covariate effect reflecting the covariate impact;
3. The error variance reflecting the unaccounted variance.

The covariate component represents the usual regression effect, and the ANOVA part represents the treatment group effect. These (treatment and covariate) do not account for all of the variation, so there is also variation in the model left unexplained and identified as *Error* in Table 6.3. The degrees of freedom serve as parameters in the F distribution. It is 1 in the covariate row as it signifies the number of covariates in the model (i.e., creatinine). The degrees of freedom are t-1 in the treatment row = the number of free parameters in that group (there are 3 groups in the example – no, yes, and metastatic) – 1. The overall mean is unknown, as it loses a degree of freedom to account for that true mean. In the error row, the degree of freedom is the difference between what is given in the total row minus all the other rows. This leads to two F-test statistics to carry out the two hypothesis tests

TABLE 6.3

Analysis of Covariance (ANCOVA)

Source	Degrees of freedom	Sums of squares	Mean square	F-test
Covariate	1	SS(cov)	MS(cov)=SS(cov)/1	MS(cov)/MSW
Treatment	t-1	SS(trt)	MSB=SSB/t-1	MSB/MSW
Error (Within)	n-t-1	SSW	MSW=SSW/n-k-1	
Total	n-1	TSS	Total	

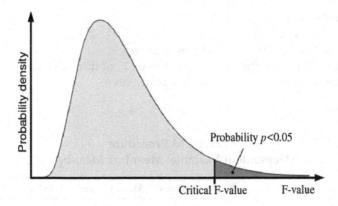

FIGURE 6.2
An example of an F distribution

within an ANCOVA model; one for the covariate effect and other for the treatment effect.

The value of that F-test statistic value in Table 6.3 provides information to make a decision about the null hypothesis. The test value gives the researcher a sense of the degree of belief or doubt about the null hypothesis. Once the F-test statistic value is obtained, then one needs to devise a way to tell when it is large enough to determine if there is sufficient evidence to support the research statement (disbelief in the null hypothesis). The calculated F-test statistic follows a F distribution (Figure 6.2). An example of an F distribution is given in Figure 6.2. This shape varies as the number of degrees of freedom in the numerator and the denominator change.

A p-value (probability that the data support the null hypothesis) is obtained based on the assumed F distribution. It gives a sense of whether one can support or disbelieve the null hypothesis. The p-value comes from the tail of F-distribution. It is the area beyond the test statistic value. The p-value needs a cut point for when it is small enough to make a decision against the null. The cut point is the critical value. The area under the F distribution, but in the tail, is the significance level or α level. It is the probability of rejecting the null hypothesis when the null hypothesis is true.

6.4 Data Analysis

6.4.1 Analysis of Data with SAS Program

The one-way ANCOVA model was fitted using PROC GLM in SAS with blood pressure as the variable of interest [MEANBP1], cancer [CA1, CA2] is the treatment variable, and creatinine [CREA1] is the covariate. As an example but not necessarily a scientific question, first, we fit one-way analysis of variance model with cancer groups.

```
data rightheart;
set rightheart;
ca1=(ca='No'); ca2=(ca='Yes');run;
proc glm data=rightheart alpha=0.1 order=data;
model meanbp1=ca1 ca2 /solution;run;
```

The SAS output gives

The GLM Procedure
Dependent Variable: Meanbp1 Meanbp1

Source	DF	Sum of Squares	Mean Square	F Value	Pr > F
Model	2	66301.360	33150.680	23.08	<.0001
Error	5732	8233906.834	1436.481		
Corrected Total	5734	8300208.194			

The total sums of squares in mean blood pressure is 8300208.19. The mean square error or the model variance is 1436.48. The F value of 23.08 with 2 and 5732 df presented a p-value of <.0001. The cancer groups have different blood pressure means.

R-Square	Coeff Var	Root MSE	Meanbp1 Mean
0.007988	48.26911	37.90093	78.52005

Since the covariate is categorical, we do not address the R-square value. The square root of the model variance is the model standard error of 37.90.

Source	DF	Type III SS	Mean Square	F Value	Pr > F
ca1	1	37516.17042	37516.17042	26.12	<.0001
ca2	1	3383.48783	3383.48783	2.36	0.1249

The F-value for ca1 has a p-value <.0001. This suggests that ca1 group is significantly different from metastatic group, and ca2 is not statistically significantly different in mean blood pressure from that of the metastatic group.

Parameter	Estimate	Standard Error	t Value	Pr > |t|
Intercept	70.05468750	1.93412381	36.22	<.0001
ca1	10.30852328	2.01714487	5.11	<.0001
ca2	3.50601209	2.28444646	1.53	0.1249

The parameter estimates are given. The intercept estimate is an overall mean value. It provides information on the overall mean as it compares to zero (t = 36.22, with p <.0001). In practice, it is not very informative. The rows for ca1 and ca2 provide theresult in comparison to metastatic group.

We fit the ANCOVA model with creatinine added in as a covariate. The SAS output gives

```
** output meanbp1(Mean Blood Pressure Day 1);
** input ca(3 levels no/yes/metastatic) & crea1(serum
            creatinine Day 1);
proc sort data = rightheart; by descending ca;
run;
data rightheart ;
set rightheart;
ca1=(ca='No'); ca2=(ca='Yes'); * metastatic was left off
            and serves as the baseline;
run;
proc glm data=rightheart alpha=0.1 order=data;
model meanbp1=ca1 ca2 crea1/solution;
run;
```

The SAS output gives

The GLM Procedure
Dependent Variable: Meanbp1

Source	DF	Sum of Squares	Mean Square	F Value	Pr > F
Model	3	139018.043	46339.348	32.54	<.0001
Error	5731	8161190.151	1424.043		
Corrected Total	5734	8300208.194			

The F-value is 32.54, with a p-value <.0001. This suggests that CREA and cancer are correlated with blood pressure.

There are 5735 observations. The total sums of squares in blood pressure is 8300208.19. Cancer ad creatinine account for 139018.043. The MSE talks about the model variance. There is a reduction in the MSE. It is now 1424.04 as opposed to MSE of 1436.481 previously. A reduction of 12.44 was brought on by creatinine. The model says it was a significant inclusion.

Source	DF	Type III SS	Mean Square	F Value	Pr > F
ca1-cancer_no	1	43143.614	43143.614	30.30	<.0001
ca2-cancer_yes	1	4767.812	4767.812	3.35	0.0673
crea1= creatinine	1	72716.683	72716.683	51.06	<.0001

The Type III SS talks about the added effect or the extra effect as it pertains to a variable in the model. It gives the same information as the parameter estimates in the model. Here the F-test is used, while there the t test is used. It reminds us that $F = t^2$ when the numerator degrees of freedom (df) for F is 1, and denominator degrees of frredom is very large.

Parameter	Estimate	Standard Error	t Value	Pr > |t|
Intercept	73.06637427	1.97131226	37.06	<.0001
ca1	11.07022037	2.01121987	5.50	<.0001
ca2	4.16531297	2.27640560	1.83	0.0673
crea1	−1.73698954	0.24307580	−7.15	<.0001

There is a statistically significant negative linear relationship between the covariate creatinine and blood pressure [p-value ≤.0001 with –1.736 as its estimate]. We find that ca1 (Cancer-no) is different from ca3 (cancer-metastatic) (p≤.0001). This difference is not seen between ca2 (cancer-yes) and ca3 (cancer-metastatic) (p = 0.0673).

6.4.2 Analysis of Data with R Program

The one-way ANCOVA model was fitted using R with blood pressure [MEANBP1], cancer [CA1, CA2], and creatinine, [CREA1].

```
model1 = aov(meanbp1~ca+crea1, data=rightheart)
summary(model1)
##          Df Sum Sq Mean Sq   F value  Pr(>F)
## ca        2  66301   33151     23.28  8.53e-11 ***
## crea1 1   72717   72717     51.06  1.01e-12 ***
## Residuals 5731 8161190 1424
```

There are 5735 observations. The total sums of squares in blood pressure is 8300208.19. Cancer and creatinine account for 139018.043. The MSE informs about the model variance.

The F-values for Cancer and Creatinine have p-values <.0001. This suggests that cancer and creatinine are statistically significantly correlated with blood pressure.

```
model2 = glm(meanbp1~ca+crea1, data=rightheart)
summary(model2)
## glm(formula = meanbp1 ~ ca + crea1, data=rightheart)
## Deviance Residuals:
##    Min     1Q      Median    3Q      Max
##    -82.92  -28.06   -15.14   34.99   178.16
## Coefficients:
##               Estimate Std. Error  t value Pr(>|t|)
## (Intercept)   73.0664   1.9713      37.065  < 2e-16  ***
## caNo          11.0702   2.0112       5.504  3.87e-08 ***
## caYes          4.1653   2.2764       1.830  0.0673 .
## crea1         -1.7370   0.2431      -7.146  1.01e-12 ***
## (Dispersion parameter for gaussian family taken to be
##               1424.043)
##     Null deviance: 8300208 on 5734 degrees of freedom
## Residual deviance: 8161190 on 5731 degrees of freedom
## AIC: 57925
## Number of Fisher Scoring iterations: 2
```

There is a statistically significant negative linear relationship between the covariate creatinine and blood pressure [p-value $\leq$.0001, with a -1.737 as its estimate]. Thus, we were justified in adjusting for creatinine before we examined the differences among the group means for cancer. We found that ca1 is different from CA-Metastatic (p$\leq$.0001). This difference is not seen between CA-Yes and CA-Metastatic (p = 0.0673).

6.5 Summary and Discussion

A one-way ANCOVA model assumes that there is a linear relationship between the mean of the subpopulation and a covariate. It assumes:

- That the slope remains the same for each level of the factor. If one adds an interaction term, then one would be allowing a different slope for each cancer group (Searle 1971). Analysis of covariance model is a combination of linear regression model and an analysis of variance model. It is the integration of a quantitative covariate and a categorical treatment variable;

- The primary difference between an ANOVA and an ANCOVA model is the addition of the continuous covariate. Such addition should be an appropriate covariate that explains a portion of the variability in the dependent variable of interest.

6.6 Exercises

Using the Right Heart Catheterization (RHC) dataset to fit a one-way ANCOVA model to Glasgow coma score with race and nuisance variable age, answer the following questions:

1. Is there a coma score differential among race?
2. Is the nuisance covariate significant?
3. How do the different categories of race compare?

References

Connors, A.F., et al.: The effectiveness of right heart catheterization in the initial care of critically ill patients. *JAMA*, 276(11), 889–897 (1996). http://doi.org/10.1001/jama.1996.03540110043030

Doncaster, C.P., Davey, A.J.H.: *Analysis of Variance and Covariance: How to Choose and Construct Models for the Life Sciences*. Cambridge University Press, Cambridge (2007)

Kirk, R.E.: *Experimental Design: Procedures for the Behavioral Sciences*. Brooks/Cole, Belmont, CA. (1968)

Muller, K.F., Fetterman, B.A.: *Regression and ANOVA*. SAS Institute, Cary, NC (2002)

Searle, S.R.: *Linear Models*. Wiley, New York (1971)

7

Statistical Modeling of Binary Outcomes with One or More Covariates: Standard Logistic Regression Model

7.1 Research Interest/Question

It is common to have research questions about binary outcomes. For example, what factors contribute to a successful operation? What are the risk factors associated with having a disease? As the variable of interest is binary, the mean of these binary outcomes is the proportion (probability). Binary response is the focus of this chapter. We introduce a standard logistic regression model to explain the mean of independent binary outcomes. Non-independent outcomes are discussed in Chapter 11.

Logistic regression models are used in applications such as classifying patients who received the flu shot or not (classification); finding factors that differentiate between someone who is obese and someone who is not obese (profiling), or predicting whether the patient will be diagnosed with Alzheimer's disease or not based on plaque score (classification). This chapter focuses on the use of logistic regression models for classification and deals with two possible responses. For cases with more than two possible outcomes, see Allison (1999), among others.

7.2 Bangladesh Data

In this chapter we use the Bangladesh survey data. The Bangladesh dataset consists of responses from more than 16,000 women taken from the 2011 Bangladesh Demographic and Health Survey (BDHS). The survey design consists of a three-level hierarchical data structure with individuals nested in 64 districts, and districts nested in seven divisions. The women in the survey were aged 13–49 and contain information on marital status (past or present) and currently non-pregnant women. The districts are the specific geographic areas within the administrative divisions, and the divisions are major cities in Bangladesh. Our binary response variable is contraceptive use

DOI: 10.1201/9781003315674-7

TABLE 7.1

Subset of Bangladesh Data

District	Age	Children	Education	Religion	Home	Radio	TV	Wealth Index	Division	CUC
1	37	3	1	1	3	0	0	3	Barisal	0
1	19	1	2	1	3	0	0	3	Barisal	0
1	25	2	2	1	3	1	0	3	Barisal	1
1	30	2	1	1	3	0	1	4	Barisal	1
1	25	2	2	1	3	0	0	2	Barisal	1
1	31	3	1	1	3	0	0	2	Barisal	1
1	23	1	2	1	3	0	0	2	Barisal	1
1	36	3	2	1	3	0	0	1	Barisal	1
1	37	3	1	1	3	0	0	1	Barisal	1

(CUC), coded as "1", and non-contraceptive use coded as "0". The information consists of current age, age at first marriage, and the number of living children. In addition, there are some binary predictors, including whether the household has radio (*Radio*), television (*Television*), and categorical predictors in highest educational level (*Education*), place of residence (*Home*), religious belief (*Religion*), and family wealth index (*Wealth Index*). A subset of the data is given in Table 7.1 for illustration.

7.3 Binary Response with Categorical Predictors and Covariates

A logistic regression model is a predictive method for binary outcomes modeling. As the outcomes are binary, such as live or dead, disease or no disease, purchase or no purchase, win or lose – to name a few – there is still a need to model the mean of the response, which is the probability, as they are binary variables. In short, binary logistic regression models are used when one wants to predict the probability of a certain outcome. However, although we make mention of cases with more than two categories elsewhere in this book, our focus in this chapter is binary outcomes. For categorical outcomes, one can read Agresti (1996), among others.

7.3.1 Models for Probability, Odds, and Logit

As the responses are binary, one cannot use a linear regression model as presented in Chapter 5. Why?

In both cases (continuous outcomes in Chapter 5 and binary outcomes in this chapter), the idea is to model the mean of the distribution of the response. In Chapter 5, it is the mean of a normal distribution (assumed to be the distribution of the responses), and we have the linear regression models. As such, the values lie between $(-\infty, +\infty)$. In this chapter, it is the mean (or the probability) of the Bernoulli distribution (assumed to be the distribution of the responses), and we have logistic regression models. As such the values are either zero or one. In linear regression, the observations can take on values between $(-\infty, +\infty)$, and the predicted value is between $(-\infty, +\infty)$. However, in logistic regression, the observations can take on either a value of 0 or 1, while the predicted value (the probability) lies between 0 and 1.

Probability: Let p_1 denote the probability of success (event) and $1 - p_1$ denote the probability of failure (*nonevent*). The basic idea is to model the probability of success (event) or the probability of failure (nonevent). As such, it is natural to want the result of the predicted value to be 0 and 1 or at least to lie between 0 and 1.

If one fitted a linear regression model with a covariate X_1 to the mean (probability) then

$$p_1 = \beta_0 + \beta_1 X_1$$

and the predicted probability (phat, $\hat{p}$)

$$\hat{p}_1 = b_0 + b_1 X_1$$

may lie outside the constraints of $0 \le \hat{p}_1 \le 1$ where b_0 and b_1 are the estimates of β_0 and β_1. For example, Figure 7.1 is a graph of $\hat{p}_1$ versus Age from a logistic regression model with some values given in Table 7.2a.

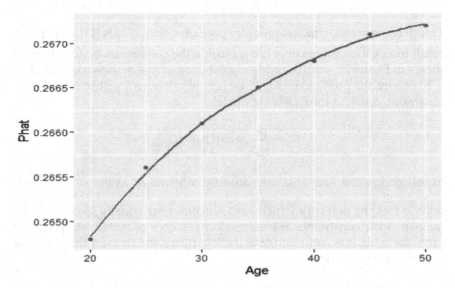

FIGURE 7.1

Phat ($\hat{p}_1$) versus Age

TABLE 7.2A

Sample Proportion by Age

AGE	20	25	30	35	40	45	50
Phat ($\hat{p}_1$)	0.2275	0.2361	0.2418	0.2458	0.2488	0.251	0.2529

It is clear that $\hat{p}_1$, the probability of using a contraceptive at a certain age is in a nonlinear relation with Age. Suppose one fits a straight line with the simple linear regression to the scatter plot; then one will expect that some of the predicted values for phat [$\hat{p}$] will be greater than 1 or less than zero for certain ages, which would be an erroneous prediction. Instead, the model of

$$p_1 = \frac{e^{\beta_0 + \beta_1 X_1}}{1 + e^{\beta_0 + \beta_1 X_1}}$$

is not linear but ensures that the predicted probabilities will never violate the constraints. While maintaining the constraints, one loses the convenience of a linear interpretation. Due to the nonlinearity formation, it is not always easy to interpret. From the probability, it is natural to think of odds.

Odds: Define the odds as $\dfrac{p_1}{1-p_1}$. In this definition, the range of the odds would be from zero to positive infinity. We can fit a linear regression to the odds as follows:

$$O = \frac{p_1}{1-p_1} = \beta_0 + \beta_1 X_1.$$

This linear regression model provides predicted odds (Odds [$\hat{O}$]) which may result in negative values thus lying outside the constraints of $0 \leq \hat{O} < \infty$, as shown in Figure 7.2, with the odds as a function of Age as shown in Table 7.2b.

To ensure the predicted odds to be non-negative and within the range of constraint, consider the following model:

$$O = \frac{p_1}{1-p_1} = \exp[\beta_0 + \beta_1 X_1].$$

In this realization, the odds are naturally constrained as $0 \leq \dfrac{p_1}{1-p_1} < \infty$ with odds = 1 as the point for which both outcomes are equally likely. While it maintains the constraints, it loses the convenience of linear interpretation. From the odds, it is natural to take the logarithm and transfer them to a linear scale, which is the logit.

Logit: The logit is the logarithm of the odds, i.e., $\text{logit}(p_1) = \log\left(\dfrac{p_1}{1-p_1}\right)$.

The logit ranges from negative infinity to positive infinity. While the odds

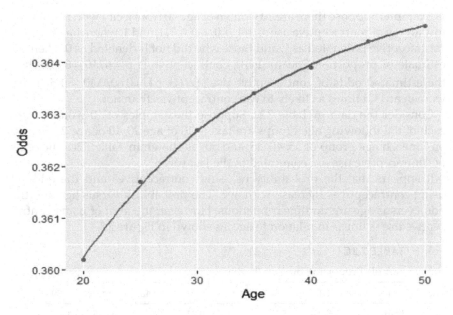

FIGURE 7.2
Odds versus Age

TABLE 7.2B

Sample Proportion by Age

AGE	20	25	30	35	40	45	50
odds	0.2946	0.3092	0.3189	0.3259	0.3311	0.3352	0.3385
Phat ($\hat{P}_1$)	0.2275	0.2361	0.2418	0.2458	0.2488	0.251	0.2529

go from 0 to infinity, the probability goes from zero to one. The logit function is useful as a means of transforming the probability to a desirable linear scale. On the logit scale, define $\log\left(\dfrac{p_1}{1-p_1}\right) = \beta_0 + \beta_1 X_1$. The logits are symmetric about zero, and they lie in the range $-\infty$ to $+\infty$. A logit with a value of 0 suggests that it is equally likely for both events and nonevents. If the identification of the two outcomes are switched, the log odds are multiplied by -1, since in the substitution $\log(p_1/(1-p_1)) = -\log((1-p_1)/p_1)$. For example, if the log odds of developing cancer are 0.05, the log odds of not developing cancer are -0.05. As the probability of an outcome increases, the odds and the log odds also increase. The log odds of an event relay the same message as the probability of the event, but on different scales. If a certain predictor has a positive impact on the logit, then it has the same positive effect on the odds and the probability, just on a different scale.

Example: Suppose there are 10 women of age 40 from a city who responded to a survey of contraceptive use as 1, 1, 0, 0, 1, 0, 0, 1, 1, and 1, where having used contraceptive is denoted as 1, and those who did not is denoted as 0. Then, an estimate of proportion (mean) using contraceptive is $\hat{p} = 6/10 = 60\%$. Then the estimated odds of contraceptive use $\left[\hat{p}/(1-\hat{p})\right] = 0.60/0.40 = 1.5$. Thus, women are 1.5 times as likely to use contraceptives than not.

Consider the data in Table 7.2c. Suppose the samples of 10 women from each of the following age groups are taken (10 of age 20, 10 of age 25, and so on for each age group of 35, 40, 45, and 50) as shown in Table 7.2c. The odds of contraceptive use are computed in the last row.

It appears that the probability of using contraceptives and the odds of using contraceptives increase with age. The probability versus age and the odds versus age are nonlinear relations. However, the logit of using contraceptive use is linear in relation to age, as shown in Figure 7.3.

TABLE 7.2C

Sample Proportion by Age

AGE	20	25	30	35	40	45	50
Logit	5.5	6.75	8	9.25	10.5	11.75	13
Odds	0.2946	0.3092	0.3189	0.3259	0.3311	0.3352	0.3385
Phat ($\hat{p}_1$)	0.2275	0.2361	0.2418	0.2458	0.2488	0.251	0.2529

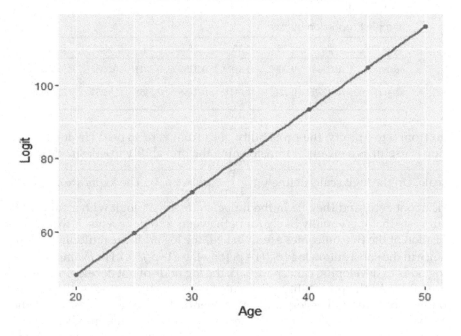

FIGURE 7.3
Logit versus Age

7.3.2 Statistical Model

The general form of the logistic regression model for the binary response Y with value one or zero given J covariates or predictors $X_1........X_J$, is

$$\text{logit}(p_1 \mid X_1........X_J) = \log\left(\frac{p_1}{1-p_1}\right) = \beta_0 + \beta_1 X_1 + ...\beta_P X_J$$

where p_1 is the probability that Y (the outcomes) is one (the event), $1-p_1$ is the probability that Y is zero (the nonevent) and $\beta_i\, i = 1,2,.....,J$ are the J unknown regression coefficients, which are estimated from the data (see Fleiss 1979). This is a mean or marginal or population-averaged model. It talks about the average rather than any subject-specific subpopulation. Although the response variable is binary (0 or 1), the logistic regression model, represented by a linear equation on the logit scale, does not predict the outcome of the binary variable but rather the probability of an outcome.

7.3.2.1 Assumptions for Standard Logistic Regression Model

When fitting a standard binary logistic regression model, there are important statistical assumptions that need to be satisfied. The observations are assumed independent, and the effect of any clustering is ignored or assumed negligible. A violation of the independence assumption may result in incorrect inferences about the regression coefficients or inefficient estimates of regression coefficients.

7.3.2.2 Interpretation of Coefficients on the Logits

The coefficients from a logistic regression equation with continuous or binary predictors are interpreted similarly to what is done in linear regression. They represent the change in logit (log odds of the response) per unit change in the

predictor. The β_j represents the change in $\log\left(\frac{p_1}{1-p_1}\right)$ with one unit change

in X_j, while the other $J-1$ covariates are held fixed. For a binary predictor, it tells how one category affects the logit versus the other category.

7.3.2.3 Interpretation of the Odds Ratio

The interpretation of the coefficients in a logistic regression model, is the odds ratio (OR). Consider two cases: One when X_j is binary, and another when X_j is continuous.

- When X_i is binary: Consider, for example, X_j as gender (female=1 and male=0) in the model. Compare two persons when everything in the covariate values is identical except for gender (male versus female) in

$$\log\left(\frac{\hat{P}_{success}}{\hat{P}_{failure}}\right) = 5 + 6X_1 + 7X_2 + 0.25 \text{ female}$$

Then, females are $e^{0.25} = 1.284$ times more likely to be successful than males;

- When X_i is continuous: Consider, for example, two persons, with one at age $X_{j(1)} = 25$ and another at age $X_{j(2)} = 26$, so there is one year difference (or consider any particular value and that particular value plus one for any of $X_{j(1)}$ and $X_{j(2)}$), therefore

$$\log\left(\frac{\hat{P}_{success}}{\hat{P}_{failure}}\right) = 5 + 6X_1 + 7X_2 + 0.25\left(Age = 25\right)$$

$$\log\left(\frac{\hat{P}_{success}}{\hat{P}_{failure}}\right) = 5 + 6X_1 + 7X_2 + 0.25\left(Age = 26\right)$$

The difference is the log odds ratio equal to 0.25. The anti-logarithmic gives a value of $e^{0.25}$. It compares two persons with the only difference being that one is one year older than the other. Thus, the older person is $e^{0.25} = 1.284$ times more likely to be successful than a person one year younger.

7.3.3 Model Fit

When measuring the fit of a model, it is customary to compare the observed values with the predicted (model) values. However, such requires one to have observed and predicted measures both on the same scale. The predicted probabilities in a logistic regression model are obtained as means on a probability scale [0, 1]. However, the observed values are either 0 or 1. The predicted is between [0, 1]. In linear regression, the observed values and the mean are on the same scale so that observed and predicted are easily compared on the scale from negative infinity to positive infinity. However, there is a dilemma in the binary case, when one attempts to compare the observed value and the model value in a logistic regression model. They are not comparable in their natural form, so they must be adjusted to be on the same scale of measurement. One approach is to dichotomize the predicted values based on 0.50 split or prior probability split, or some other appropriate cut point. While another approach is to rely on a graph of sensitivity versus 1-specificity (see Chapter 2) for the use of ROC-curves. We summarize these approaches as follows:

7.3.3.1 Classification Table

One hundred respondents were interviewed regarding their contraceptive use. A logistic regression model was fitted. A classification table based on the matching (observed, predicted) is provided. A particular observation with a "1" as observed and 0.86 as predicted, then with a cut point of 0.50, which is less than 0.86, then that point is classified as [1, 1]. However, if the outcome is 0 and the predicted is "0.34" then the point is classified as [0, 0]. A similar coding will result in misclassification in (1, 0) and (0, 1) as demonstrated in Table 7.3a. There are 85 (= 50+35) correctly classified. There are 10 as (1, 0) and 5 (0, 1), which are misclassified.

Then the percent correct is (50+35)/(50+10+5+35) = 85%. There are 15% misclassified by this model. For our example, in modeling contraceptive use with 16,186 women, the following classification table was obtained, shown in Table 7.3b.

There are (8754+1528)/16186 = 63.5% that were in agreement between the observed and the dichotomized predicted. There are 36.5% [= (4980+924)/16186] misclassified. What can we do with this information? How good is the fit?

7.3.3.2 Hosmer-Lemeshow Test – Measure of Fit

The Hosmer-Lemeshow test statistic tests the fit of the predictions from the model and the observed data. The statistic is constructed based on an ordering of the predicted probabilities then dividing into 10 (usually) groups of equal or near-equal size. A comparison of predicted and observed results in a 2 × 10 contingency table are shown in Table 7.4.

A chi-square statistic is used to check the fit of the model for the data in a contingency table. A nonsignificant chi-square test indicates that you cannot reject that the model fits the data (Hosmer and Lemeshow 2000), although there are some known issues using this test. This test is very dependent on sample size, so the value cannot be interpreted in isolation from the size of the sample (Hosmer and Hjort 2002).

TABLE 7.3A

Classification Table

	Observed	Predicted
	Use contraceptive =1	Do not use=0
Use contraceptive =1	50	10
Do not use =0	5	35

TABLE 7.3B

Classification Table

	Observed	Predicted
	Do not use =0	
Use contraceptive =1	8754	924
Do not use =0	4980	1528

TABLE 7.4

Partition for the Hosmer and Lemeshow Test

Group	Total	CUC = 1		CUC = 0	
		Observed	Expected	Observed	Expected
1	1625	531	632.76	1094	992.24
2	1618	790	805.23	828	812.77
3	1619	899	881.00	720	737.00
4	1631	911	935.70	720	695.30
5	1641	1065	981.61	576	659.39
6	1607	1072	996.43	535	610.57
7	1620	1119	1039.44	501	580.56
8	1620	1132	1078.01	488	541.99
9	1620	1118	1127.84	502	492.16
10	1585	1041	1199.99	544	385.01

7.3.3.3 ROC Curves

This use of ROC curve is applicable to situations where one has to make a decision on a model for binary responses. For example, at the airport, it is in Transportation Security Administration's (TSA) interest to correctly classify passengers as having a gun or not. These classifications must be reliable so special methods for making these decisions are used. ROC curves are used to evaluate how well these methods perform. The aim with ROC curves is to determine the area under the curve. It is the measure that determines how well the model performs. The larger the area, the better the fit.

The ROC curve represents a plot of the sensitivity (the proportion of individuals with the event who are correctly detected by the test) versus (1-specificity (the proportion of individuals who are correctly identified as not having the event)) (Chapter 2). As demonstrated in Figure 7.4, the more the curve extends to the upper left vertex, the better is the fit.

7.4 Statistical Analysis of Data with SAS and R

The decision is to use a binary logistic regression model to explain CUC with covariates age, television (TV), religion, and number of children. These variables are summarized in Table 7.5. Age is continuous, education is categorical, and TV and children are binary.

A standard logistic regression model using the statistical packages, SAS and R, programs are summarized in this section. The fitted binary logistic regression model based on these data is:

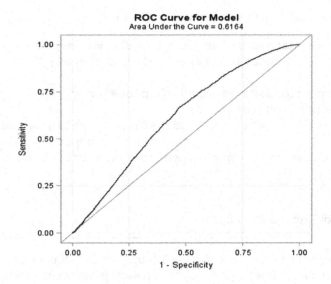

FIGURE 7.4
Example of ROC curve

TABLE 7.5

Summary Statistics for Variables

Variable	Mean	Std Dev	Minimum	Maximum	N
CUC	0.598	0.490	0	1	16186
Age	31.421	9.191	13	49	16186
TV	0.457	0.498	0	1	16186
Religion	1.123	0.359	1	4	16186
Children	2.402	1.590	0	10	16186

$$\log\left[\frac{\hat{P}_{cuc=1}}{\hat{P}_{cuc=0}}\right] = 0.56 - 0.048age + \begin{cases} +0.182 \\ +0.182 \\ +0.673 \end{cases} education + 0.135tv$$
$$+ 0.330religion + 0.317children.$$

This suggests that older people were less likely to use contraceptives (negative relation). Those who had religious beliefs were more likely to use contraceptives, and those who had more children were likely to use contraceptives. As for education (a categorical variable), there are four categories ranging from little education at 0, 1, 2, to more education at 3. One category is used as the reference. In this presentation, it was education at level 0. The model tells that the more education a woman had acquired, the more likely she is to use contraceptives. If one uses a different category as the reference, the estimates would be different numerically, but the interpretation is the same.

7.4.1 Statistical Analysis of Data with SAS Program

```
proc means data=Bangladesh mean std min max N;
var CUC age education TV religion children;
run;
proc logistic data=Bangladesh plots(only)=ROC;
class education(ref="0");
model CUC(event="1")=age education TV religion children/
                rsq lackfit Influence;
output out=outreg1 pred=predict h=leverage resdev=dev
                reschi=chi;
run;
```

The SAS output consists of:

In SAS, the ROC curve is obtained through "=ROC". Rather than repeat the curve four times, one for each statistical program, we presented it earlier in Figure 7.4.

The LOGISTIC Procedure

Model Information

Dataset	WORK. BANGLADESH	
Response variable	CUC	CUC
Number of response levels	2	
Model	binary logit	
Optimization technique	Fisher's scoring	
Number of observations read		16186
Number of observations used		16186

The model is referred to as the binary logit (or refer to as binary logistic) model. To obtain the beta estimates, we use an optimization technique in Fisher's scoring.

Response Profile

Ordered Value	CUC	Total Frequency
1	0	6508
2	1	9678

Probability modeled is CUC=1.

If fit logit [CUC=1] then it is $log\left[\dfrac{P_1}{P_0}\right]$ and not $log\left[\dfrac{P_0}{P_1}\right]$.

Class Level Information

Class	Value	Design Variables		
Education	0	−1	−1	−1
	1	1	0	0
	2	0	1	0

The design matrix is given for categorical variables (education in this case). To obtain unbiased estimates, we must reparametrize, using one less than the number of categories. We can reparametrize in different ways. One example is to let one category take on the value 0.

There are times that these models will not converge, especially when dealing with correlated data. This note tells us there is convergence.

Model Convergence Status

Convergence criterion (GCONV=1E−8) satisfied.

Model Fit Statistics

Criterion	Intercept Only	Intercept and Covariates
AIC	21815.690	21116.455
SC	21823.382	21177.990
−2 Log L	21813.690	21100.455
R-Square	0.0431	Max-rescaled R-Square	0.0582

These statistics provide a measure of the model fit. There are no p-values attached, however, to talk about how good a fit it is. You can use them to compare two nested models. However, the difference for intercept only and intercept and covariates for −2LogL is 21813.690−2110.455 = 713.2353, which is the likelihood ratio for testing the covariates in the model. The covariates are statistically significant. Since the outcome variable is binary, we do not attempt to interpret R-square.

Testing Global Null Hypothesis: BETA=0

Test	Chi-Square	DF	Pr > ChiSq
Likelihood ratio	713.2353	7	<.0001
Score	700.3984	7	<.0001
Wald	660.2711	7	<.0001

Testing that age, education, TV, religion, and children together impact contraceptive use is tested with any one of these three test statistics: likelihood ratio, score, or Wald. (p <.0001). It implies that at least one of these covariates is correlated with contraceptive use. This addresses all the covariates together as a group.

	Type 3 Analysis of Effects		
Effect	**DF**	**Wald Chi-Square**	**Pr > ChiSq**
age	1	410.7138	<.0001
education	3	84.3150	<.0001
TV	1	14.4331	0.0001
religion	1	47.7501	<.0001
children	1	493.8406	<.0001

This table shows that age, education, TV, religion, and children are each statistically significant in the model. p-value [Pr > ChiSq] < .0001.

				Standard	**Wald**	
Parameter		**DF**	**Estimate**	**Error**	**Chi-Square**	**Pr > ChiSq**
Intercept		1	0.8197	0.0808	102.9969	<.0001
Age		1	−0.0482	0.00238	410.7138	<.0001
Education	1	1	−0.0772	0.0298	6.7034	0.0096
Education	2	1	−0.0773	0.0291	7.0351	0.0080
Education	3	1	0.4136	0.0495	69.8652	<.0001
TV		1	0.1350	0.0355	14.4331	0.0001
Religion		1	0.3303	0.0478	47.7501	<.0001
Children		1	0.3171	0.0143	493.8406	<.0001

(Analysis of Maximum Likelihood Estimates)

The logistic regression model is

$$\log\left[\frac{\hat{P}_{cuc=1}}{\hat{P}_{cuc=0}}\right] = 0.82 - 0.05age + \begin{Bmatrix} -0.08 \\ -0.08 \\ +0.41 \end{Bmatrix} education + 0.135tv + 0.33religion + 0.32children$$

	Odds Ratio Estimates		
Effect	Point Estimate	95% Wald Confidence Limits	
Age	0.953	0.949	0.957
Education 1 vs 0	1.200	1.100	1.309
Education 2 vs 0	1.200	1.093	1.316
Education 3 vs 0	1.960	1.696	2.265
TV	1.144	1.068	1.227
Religion	1.391	1.267	1.528
Children	1.373	1.335	1.412

Taking the exponent of the estimates gives the odds ratio. For example, TV has an estimate of 0.1350, so a person having a TV is $e^{0.1350} = 1.144$ times more likely to use contraceptives. These data suggest that younger people, people with a higher level of education, people with a TV in the home, people with religious beliefs, and people with more children are more likely to use contraceptives.

Hosmer and Lemeshow Goodness-of-Fit Test		
Chi-Square	DF	Pr > ChiSq
174.5039	8	<.0001

The Hosmer-Lemeshow goodness of fit test suggests that the model does not fit the data [p <.0001]. For goodness of fit, the null hypothesis implies the model fits the data. With such a p-value, one declares that the model does not fit the data.

Case Number	CUC	Predicted Values	Cook's D	Pearson Residual	Deviance Residual	Hat Matrix Diagonal
1	0	0.5603	0.000390	−1.1288	−1.2819	0.000306
2	0	0.6165	0.000533	−1.2680	−1.3846	0.000331
3	1	0.6232	0.000177	0.7776	0.9726	0.000293
4	1	0.5980	0.000238	0.8198	1.0140	0.000355
5	1	0.6232	0.000177	0.7776	0.9726	0.000293
6	1	0.6298	0.000152	0.7667	0.9617	0.000258
7	1	0.5701	0.000240	0.8684	1.0601	0.000318
8	1	0.5721	0.000298	0.8649	1.0568	0.000399
9	1	0.5603	0.000240	0.8859	1.0764	0.000306
10	1	0.6143	0.000177	0.7924	0.9872	0.000282
11	0	0.7143	0.000797	−1.5813	−1.5830	0.000319

The truncated table provides predicted values, influential measures (Cook's D), Pearson residuals, deviance residuals, and leverages (hat matrix). The predicted counts for the first observation, Pred = 0.560, with Pearson and deviance residual (we use standardized residuals to identify outliers) both between [−1.96, +1.96] and suggest that it is not an outlier due to response. The hat matrix value or leverage tells whether it is not an outlier due to covariate. Whenever the logistic regression analysis encounters influential observations, one must investigate the issue and take the appropriate actions. However, observations may be classified as influential even though they are not recognized as outliers.

7.4.2 Statistical Analysis of Data with R Program

```
## Call "glm" to fit logistic regression
glm.out = glm(CUC~age+factor(education)+TV+religion+children,
              family=binomial((link='logit')), data=Bangladesh)
```

```
summary(glm.out)
## glm(formula = CUC ~ age + factor(education) + TV + religion +
##     children, family = binomial((link = "logit")), data =
              Bangladesh)
## Deviance Residuals:
##       Min        1Q     Median        3Q       Max
##    -2.2319   -1.2661    0.8506    1.0023    1.8094
## Coefficients:
##                    Estimate  Std. Error z value Pr(>|z|)
## (Intercept)        0.560506    0.090299   6.207 5.39e-10 ***
## age               -0.048154    0.002376 -20.266  < 2e-16 ***
## factor(education)1 0.182064    0.044442   4.097 4.19e-05 ***
## factor(education)2 0.181950    0.047462   3.834 0.000126 ***
## factor(education)3 0.672870    0.073815   9.116  < 2e-16 ***
## TV                 0.134967    0.035526   3.799 0.000145 ***
## religion           0.330278    0.047796   6.910 4.84e-12 ***
## children           0.317067    0.014268  22.223  < 2e-16 ***
```

The logistic regression model is

$$\log\left[\frac{\hat{P}_{cuc=1}}{\hat{P}_{cuc=0}}\right] = 0.561 - 0.048age + \begin{cases} +0.182 \\ +0.182 \\ +0.673 \end{cases} education$$
$$+ 0.135tv + 0.330religion + 0.317children$$

Taking the exponents of the estimates gives the odds ratios. For example, TV has an estimate of 0.1350, so a person having a TV is $e^{0.1350} = 1.144$ more likely to use a contraceptive. These data suggest that younger people, people with a higher level of education, people with a TV in the home, people with religious beliefs, and people with more children are more likely to use contraceptives.

```
## (Dispersion parameter for binomial family taken to be 1)
##      Null deviance: 21814 on 16185 degrees of freedom
## Residual deviance: 21100 on 16178 degrees of freedom
## AIC: 21116
## Number of Fisher Scoring iterations: 4
predict<-glm.out$fitted.values      # predicted value
cooksD<-cooks.distance(glm.out)     # Cook's D
leverage<-hatvalues(glm.out)        # Leverage
residuals<-resid(glm.out)           # Residuals
rstudent<-rstudent(glm.out)         # compute Student Residuals
out<-data.frame(CUC,predict,cooksD,leverage,rstudent,residuals)
names(out)=c('CUC','Phat', 'CooksD', 'leverage','StudResid','D
                eviance')
print(round(out,digits=6))
```

```
## CUC      Phat     CooksD    leverage  StudResid Deviance
## 1    0   0.560290  0.000049  0.000306 -1.282056 -1.281904
## 2    0   0.616536  0.000067  0.000331 -1.384756 -1.384564
## 3    1   0.623166  0.000022  0.000293  0.972656  0.972565
## 4    1   0.598043  0.000030  0.000355  1.014112  1.013994
## 10   1   0.614288  0.000022  0.000282  0.987299  0.987209
```

The table provides predicted values, influential measures (Cook's D), standardized residuals, deviance residuals, and leverages. The predicted counts for the first observation, Phat = 0.560, with standardized residual and deviance residual between [−1.96, +1.96] suggest that it is not an outlier due to response. The leverage tells whether it is not an outlier due to covariate. Whenever the logistic regression analysis encounters influential observations, one must investigate the issue and take the appropriate actions. However, observations may be classified as influential even though they are not recognized as outliers.

7.5 Research/Questions and Comments

A standard logistic regression model is a marginal or population-averaged model. It models the mean of a binary outcome variable. It also identifies the most likely characteristic of candidate for certain outcomes. It is an important tool for classification and profiling.

The Hosmer-Lemeshow test at times may reveal that the standard logistic regression is not a good fit, though the covariates are significant. In such cases, this is not particularly useful, as the model is not a good fit. In our experience, these data assume that there are independent observations. However, these data came from a hierarchical structure. As such, the observations are correlated, so the variance is most likely to be larger than what the standard logistic regression is estimating it to be. In Chapter 11, we will return to the analysis of these data with more models to examine a structure for hierarchical data.

7.6 Questions

1. What should be done if the Hosmer-Lemeshow-test was rejected?
2. If we ignore the overdispersion that is present due to the correlated data, what are the consequences?

Answers

1. Fit a GEE model or a generalized linear mixed model.
2. The fact that the overdispersion was ignored results in underestimation of the true standard error. Thus, the test statistics are larger than what they really are, resulting in smaller p-values. As such, one is likely to declare significance when there is not.

7.7 Exercises

1. Fit a logistic regression model to the contraceptive use data for Division 1 using the covariates age, education, TV, religion, and children.

2. Is the model a good fit?

3. How do the results differ when applied to all divisions?

4. Who is most likely to use contraceptives in Division 1?

References

Agresti, A.: *An Introduction to Categorical Data Analysis*. Wiley, New York (1996)

Allison, P.: *Logistic Regression Using the SAS System: Theory and Application*. SAS Institute, Cary, NC (1999)

Fleiss, J.L.: Confidence intervals for the odds ratio in case-control studies: The state of the art. *Journal of Chronical Diseases*, 32, 69–77 (1979)

Hosmer, D.W., Hjort, N.L.: Goodness-of-fit processes for logistic regression: Simulation results. *Statistics in Medicine*, 21, 2723–2738 (2002)

Hosmer, Jr, D.W., Lemeshow, S.: *Applied Logistic Regression*, 2nd ed. Wiley, New York (2000)

8

Generalized Linear Models

8.1 Research/Question

In research and in fit of models, one is often confronted with modeling the mean of certain random responses with a cadre of possible predictors. These predictors may or may not have an effect on the outcome. The responses may or may not come from a normally distributed population. This chapter provides an overview of models that have responses that do not necessarily come from a normal distribution but is a member of the exponential family. These models belong to a class of generalized linear models (GLMs). A GLM consists of three components: the random component, the systematic component, and the link component.

In deciding on the appropriate model of the generalized linear model, we must examine the type of response and the associated distribution. It is not uncommon to think of the outcomes as coming from a normal distribution or a transformation of the outcome as a normal population. While the transformation may be appropriate, the interpretation, at times, is challenging, as one needs the original scale to present the interpretation.

This chapter provides several examples to identify the use of generalized linear models. Models with binary and continuous outcomes are examined.

8.2 German Breast Cancer Study Data

The German Breast Cancer Study Group recruited 720 patients with primary node-positive breast cancer into the Comprehensive Cohort Study (Schmoor, Olschhewski, and Schumacher 1996). Both randomized and nonrandomized patients were eligible, and about two-thirds were entered into the randomized part to examine the effectiveness of three versus six cycles of chemotherapy as well as the hormonal treatment with tamoxifen. A follow-up lasting about five years resulted in 312 patients who had at least one recurrence of the disease or had died. These data were fitted in Chapter 3 without covariates. In this chapter, these data are revisited to demonstrate statistical models for cases with continuous outcome measures and binary outcome measures but without considering any differential effects due to any covariates (Sauerbrei and Royston 1999). A subset of the data follows in Table 8.1. In Table 8.1, there are age, menopause, size, nodes, and use of tamoxifen.

8.3 Generalized Linear Model

8.3.1 The Model

The term *generalized linear model* refers to a larger class of models. It includes general linear models and linear models (McCullagh and Nelder 1989). Generalized linear models provide a means of modeling the mean parameter of the distribution with relationship while adjusting for one or more covariates. However, the outcome may arise on account of any assumed distribution in the exponential family.

In probability and statistics, the exponential family is an important class of probability distributions that share a specific structural form with certain fixed parameters presented in terms of exponentials. The concept of exponential families was credited to E. J. G. Pitman, G. Darmois, and B. O. Koopman in 1935–1936. The exponential families include many of the most common distributions, including the normal, exponential, gamma, chi-square, beta, Dirichlet, Bernoulli, Poisson, Wishart, inverse Wishart, and many others. Distributions are members of the exponential family only when certain parameters are considered fixed and known (e.g., binomial, multinomial, and negative binomial). Examples of common distributions that are not exponential families are Student's t, most mixture distributions, and the family of uniform distributions with unknown bounds.

The variable of interest (the response variable) is often referred to as outcome variable, target variable, or the so-called Y variable. If the response always resulted in the same outcome, then there is nothing to model. A GLM model explains the mean of the outcomes through the levels of the covariates.

TABLE 8.1

Subset of Breast Cancer Data

id	age	meno	size	nodes	hormon
132	49	premenopausal	18	2	no tamoxifen
1575	55	Postmenopausal	20	16	no tamoxifen
1140	56	Postmenopausal	40	3	no tamoxifen
769	45	premenopausal	25	1	no tamoxifen
130	65	Postmenopausal	30	5	had tamoxifen
1642	48	premenopausal	52	11	no tamoxifen
475	48	premenopausal	21	8	no tamoxifen
973	37	premenopausal	20	9	had tamoxifen

The covariates comprise the systematic part of the model. The error term accounts for the random part of the model. It represents the unexplained variation but is believed to follow some distributional random pattern, as discussed in Gary (2014) and Anderson et al. (2007). Once the outcomes and the scale of measurements are known, one can suggest possible types of relationships between the set of covariates and the mean of the outcome.

Once the model is fitted, it is necessary to know how well the data support a belief in a result, (a relation between the covariates and the outcomes). This is achieved by examining the impact of the covariates and interpreting the effect on the outcomes.

There are three components to any GLM.

The *random component* represents the distribution of the outcome. Within the random component lies the distribution of the outcomes. This is the same as referring to the distribution of the unmeasurable effects. In linear regression models, the random component distribution is assumed to be normal. But it can be, for example, the Poisson distribution, which implies that outcome is a count. In fact, the random component of GLMs can follow any distribution from a family of distributions referred to as the exponential family of distributions. The random component addresses the left side of the model equation through the distribution of the outcome.

The *systematic component* refers to the portion of the variation in the outcomes that is explained by some combination of the covariates. The systematic component represents the right side and is consider fixed. The covariates are combined through coefficients, $\beta_0, \ldots, \beta_p$ and addresses the right side of the model equation.

The *link function* connects the systematic and random components. The link function, as the name implies, links the two components (random and systematic), thereby presenting the generalized linear model. The user gets to specify the link function, although there are common link functions typically selected based on the distribution in the random component.

In summary, a generalized linear model consists of

$$\text{Twice differentaible function}\left[\text{mean of distribution Outcome}_i\right]$$
$$= \text{Linear combination}\left[\text{of parameter and covariate}\right].$$

Generalized linear model is a broad class of models that include linear regression, analysis of variance (ANOVA), logistic regression, Poisson regression, log-linear models, etc. A summary of the generalized linear model, as presented in Agresti (2013), is reproduced in Table 8.2. The table gives an overview of examples of generalized linear models while presenting their components (Burnham and Anderson 1998). For example, in the logistic regression model in Chapter 8, its random component is a Bernoulli distribution, the systematic component contains covariates, which can be continuous or categorical, and the link function is the logit function.

TABLE 8.2

Link Components for Models

Model	Random	Link	Systematic
One-sample t-test	normal	identity	constant
Two-sample t-test	normal	identity	constant
Linear regression	normal	identity	continuous
ANOVA	normal	identity	categorical
ANCOVA	normal	identity	mixed*
Logistic regression	binomial	logit	mixed*
Loglinear	Poisson	log	categorical
Poisson regression	Poisson	log	mixed*
Multinomial response	multinomial	generalized logit	mixed*

Mixed indicates continuous and categorical predictors.

$$\text{logit}(probability) = \beta_0 + \beta_1 X_1 + \ldots\ldots + \beta_p X_p$$

8.3.2 Assumptions When Fitting GLMs

When fitting generalized linear models, the assumptions are:

Assumption 1: Instead of assuming that the outcomes are from a normal distribution, one assumes that the outcome comes from a distribution that belongs to the exponential family of distributions. (This includes the well-known normal distribution (one parameter), binomial distribution, and Poisson distribution, etc.);

Assumption 2: The mechanism that gives rise to an outcome is not related to the mechanism that gives rise to other outcomes (the observations are independent);

Assumption 3: A linear relationship exists between a function of the mean parameter of the distribution and the covariates. Since the interest is a function of the mean, these are marginal models or population-averaged models. For example, in modeling counts data, a Poisson regression model is often used. It belongs to the class of GLMs with the following three components:

In the Poisson regression model

1. The outcomes are the observed counts. Assume the outcomes are independent observations. Assume the outcomes from a Poisson distribution (*random component*);
2. A linear combination consists of the covariate $\beta_0 + \beta_1 X_1 + \ldots\ldots + \beta_p X_p$ (*systematic component*);
3. The function, which links a linear combination of the predictors and the mean of the response variable is log (*link component*).

Therefore, in Poisson regression models, the counts are outcomes from the Poisson distribution with mean λ and variance λ, thus our model is

$$\log(\lambda) = \beta_0 + \beta_1 X_1 + \ldots\ldots + \beta_p X_p.$$

Essentially, a generalized linear model describes how a function of the mean relates linearly to the set of predictors. In particular, the fitted model is the log of the estimator λ, such that

$$\log(\hat{\lambda}) = \hat{\beta}_0 + \hat{\beta}_1 X_1 + \ldots\ldots + \hat{\beta}_p X_p,$$

where $\hat{\lambda}$ denotes an estimate of the mean of the random variable from the outcome. The effect of a unit change in X_i changes the $\log(\hat{\lambda})$ in an amount $\hat{\beta}_i$ (an estimate of β_i, which is the ith element in the set of parameters). Equivalently, the effect of a unit change in X_i leads to an increase in the log of a positive response multiplicatively by the factor $\exp(\beta_i)$, while other predictors are held fixed.

The generalized linear models (see McCullagh and Nelder 1989) provides an extension of linear models, allowing one to use other distributions than the normal. It negates the need for the assumptions of normality, constant error variance, and a linear relationship between the covariate effects and the mean.

8.4 Examples of Generalized Linear Models

8.4.1 Multiple Linear Regression

A multiple linear regression model tells how the mean of a continuous normal response variable is driven by a set of covariates. Chapter 5 presented a multiple linear regression model with response BMI and predictors, age, waist, and albumin. This multiple linear regression model is a member of the class of generalized linear model shown as follows:

1. *Random component: $Y = [BMI]$* is a continuous response or outcome variable that follows a normal distribution at each subpopulation. The desire is to explain the mean of BMI, $\mu_{Y|X_1, X_2, X_3}$;

2. *Systematic component: the linear combination of covariates* (continuous or categorical)

$$\beta_0 + \beta_1 \text{Age} + \beta_2 \text{Waist} + \beta_3 \text{Albumin}$$

that is linear in the parameters $\beta_0, \beta_1, \beta_2$, and β_3;

3. *Link component: identity link I, which models the mean directly related,*

therefore, the model is:

$$I(\text{mean of BMI}) = \mu_{Y|X_1, X_2, X_3} = \beta_0 + \beta_1\text{Age} + \beta_2\text{Waist} + \beta_3\text{Albumin}.$$

8.4.2 Logistic Regression

Logistic regression models represent the mean of a binary response/outcome variable Y and depends on a set of P covariates, $X_1, \ldots, X_P$. It addresses the logit (log odds of an event occurring) as a function of the probability. The occurrence of the event is usually measured as "1" and the nonoccurrence (nonevent) is measured as "0". In Chapter 7, a fit of a logistic regression model with response to contraceptive use with predictors of age, education, television use, religion, and the number of children is given. A demonstration of its membership in the class of generalized linear models follows:

The logistic regression model has binomial error (whether the subject uses contraceptive or not) and a logit link [$\text{logit}(P_{AD})$] (Harrell 2001). In other words, while there is no linearity on the probability scale, there is on the logit scale.

1. *Random component:* The distribution of $Y = [Contraceptive\,Use]$ is Bernoulli, where P_{CUC} is the probability of contraceptive use. The mean is the proportion, P_{CUC};

2. *Systematic component:* with *covariates* age, education, television use, religion and number of children is

   $$\beta_0 + \beta_1\text{Age} + \beta_2\text{Educ} + \beta_1\text{TV} + \beta_3\text{Use} + \beta_4\text{Religion} + \beta_5\text{Children}$$

   which is linear in the parameters $\beta_1, \beta_2, \beta_3, \beta_4$ and β_5;

3. *Link function:* In logistic regression, the link function is the logit link, which is the log of the odds or $\log\left(\dfrac{P_{CUC}}{P_{NoCUC}}\right)$, where P_{NoCUC} denotes the probability that the woman does not use a contraceptive. The odds are the ratio of two probabilities, event versus nonevent;

4. Model:

$$\log\left(\frac{P_{CUC}}{P_{NoCUC}}\right) = \beta_0 + \beta_1\text{Age} + \beta_2\text{Educ} + \beta_1\text{TV} + \beta_3\text{Use} + \beta_4\text{Religion} + \beta_5\text{Children}.$$

8.5 Numerical Poisson Regression Example

In this section, we outline the fit of a generalized linear model using SAS and R statistical software. Consider predicting the mean of nodes using a log link. The responses satisfy the requirements for the random component to be considered Poisson. A summary of the data, consisting of the mean, range, minimum, maximum, standard deviation, variance, and standard deviation of the mean (standard error) follows in Table 8.3:

Poisson regression is a member of a class of GLM where the random component is the Poisson distribution for the response variable, a count variable. When all explanatory variables are discrete, the Poisson regression is equivalent to the log-linear model. For example, the German Breast Cancer Study Group recruited 720 patients with primary node-positive breast cancer into the Comprehensive Cohort Study (Schmoor et al. 1996) in which we fit a Poisson regression model.

1. *Random component:* The number of nodes (count) is the response variable, which is distributed as a *Poisson*;
2. *Systematic component:* A linear combination of the covariates are $\beta_1 \text{Size} + \beta_2 \text{Age} + \beta_3 \text{Hormon}$ and are linear in the parameters β_1, β_2 and β_3;
3. *Link function:* The data are modeled on the logarithmic scale, with a log link.

Model:

$$\log(\text{cell count}) = \beta_0 + \beta_1 \text{Size} + \beta_2 \text{Age} + \beta_3 \text{Hormon}$$

The fitted model using any of the two statistical programs is

$$\log[nodes] = 0.71 + 0.019size + 0.005age + 0.026hormon$$

TABLE 8.3

Descriptive Statistics for Variables in Breast Cancer Data

	N	Minimum	Maximum	Mean	Std. Deviation
Age	686	21	80	53.05	10.121
Nodes	686	1	51	5.01	5.475
Size	686	3	120	29.33	14.296
Hormon	686	0	1	.36	

Interpretation of parameter estimates:

- The effect on the mean of the nodes is $e^{0.71} = 2.03$, thus it is a measure of the effect on the mean nodes when age is zero, size is zero, and when there is no tamoxifen. Though this may not have practical interpretation, it sets the stage for such interpretations;
- For every unit increase in size, we expect log (nodes) to increase by 0.019. As $e^{0.019} = 1.0192$, it means that we expect the mean nodes to increase by 1.019 for every unit increase in size, while other covariates are held fixed;
- For every unit increase in age with $e^{0.005} = 1.005$, nodes increase by 1.005 times, while other covariates are held fixed. If there is tamoxifen, expect the increase in the mean nodes to be 1.026 over no tamoxifen. The predictors have a multiplicative effect on the mean nodes;
- In particular, suppose there is a coefficient equal to 0, then $e^{0.00} = 1.00$, and the expected count is not related to that predictor. If the coefficients were negative, then $e^{-ve} < 1.00$, and the expected count is less than the overall mean.

Does the model fit well? Obtain a ratio of the goodness-of-fit statistic (deviance) to its degrees of freedom (= scale factor). If the ratio is larger than the value of one, given that the statistic was significant, then it signifies that model is not a good fit. A lack of fit may be due to missing data, covariates left out, or overdispersion. The predictor is significant, but the model does not fit well.

How does one try to adjust for overdispersion to see if one can improve the model fit? Recall that one of the reasons for overdispersion is the heterogeneity, where subjects within each covariate combination still differ greatly. If that is the case, which assumption of the Poisson model is violated?

Goodness of Fit			
	Value	df	Value/df
Deviance	2611.739	682	3.830
Scaled deviance	2611.739	682	
Pearson chi-square	3393.728	682	4.976
Scaled Pearson chi-square	3393.728	682	
Log likelihood[b]	−2360.113		
Akaike's Information Criterion (AIC)	4728.227		
Finite Sample Corrected AIC (AICC)	4728.286		
Bayesian Information Criterion (BIC)	4746.350		
Consistent AIC (CAIC)	4750.350		

Dependent variable: nodes; Model: (Intercept), age, size, hormone.
[a] Information criteria are in smaller-is-better form.
[b] The full log likelihood function is displayed and used in computing information criteria.

The ratio ($\frac{value}{df}$ = 3.830 and 4.976)is much larger than one. This suggests that there is overdispersion. One can use the scale to adjust the variance and refit the model.

			Parameter Estimates				
			95% Wald Confidence Interval		Hypothesis Test		
Parameter	B	Std. Error	Lower	Upper	Wald Chi-Square	df	Sig.
(Intercept)	.713	.1017	.513	.912	49.125	1	.000
age	.005	.0018	.002	.009	8.886	1	.003
size	.019	.0009	.017	.021	447.701	1	.000
hormon	.026	.0365	−.046	.097	.503	1	.478
(Scale)	1[a]						

Dependent Variable: nodes; Model: (Intercept), age, size, hormone;
[a] Fixed at the displayed value.

Predicted, Leverage, Residual, Standardized Residual, Cook's Distance

The predicted counts for the first observation, Pred = 3.810, and log (3.810) = 1.3377, so residual is −1.735. Whenever the GLM determines an observation as influential, one must investigate the issue and take appropriate actions. One uses standardized residuals to identify outliers. However, observations may be classified as outliers even though they are not recognized as influential (Agresti 2007). A summary of a four-step process of identifying outliers, leverage points, and influential observations is presented below.

Step 1: Residuals (difference between the observation and the model value) play a role in identifying observations that are outliers due to the magnitude of the response in relation to others. The residual is the primary means of classifying an observation as an outlier. It is best to standardize the residual. To standardize the residual, one takes the residuals and divides each one by its standard deviation. As such, the standardized results follow a z distribution. Observations falling outside the [−2, +2] are statistically significant in their difference from 0 and are considered outliers. This means that the predicted value is also significantly different from the actual value;

Step 2: Identify those observations that are substantially different from the remaining observations based on the covariate's values. Such cases are determined using the leverage points. These are values obtained from the so-called "hat matrix". The hat matrix values represent the combined effects of all covariates for each case. The range of possible values of the hat matrix goes from 0 to 1. The average value is p/n, where p is the number of predictors (the number of coefficients plus one for the constant) and n is the sample size;

Step 3: Capture the impact of an observation from two sources: the size of changes in the predicted values if the case is omitted (outlying standardized residuals); and the observation's distance from the other observations (leverage). Cook's distance is the single most representative measure of influence on overall fit. Large values (usually greater than 1) indicate substantial influence by the case in affecting the estimated regression coefficients;

Step 4: Identify all observations exceeding the threshold values and then examine the dataset and look for outstanding values. Some additional observations will be detected that should be classified as influential. The researcher has to be careful when deciding to eliminate them. There are always outliers in any population, and the researcher must be careful not to trim the dataset, so that good results are almost guaranteed, as shown in Table 8.4.

Table 8.4 provides the predicted, leverage, residual, standardized residual, and Cook's values. In Table 8.4, id#1575 has large standardized values. Such an observation is not consistent with the model.

Change the Model: Adjusting for Overdispersion

If the scale factor is greater than 1, then there is a potential problem with over-dispersion. It tells that the mean is not equal to the variance, as is expected in a Poisson distribution for the number of nodes. If one wants to test and

TABLE 8.4

Predicted, Leverage, Residual, Standardized Residual, Cook's

id	Mean Predicted	Leverage	Residual	Pearson Residual	Deviance Residual	Std Pearson Residual	Cooks Distance
132	3.735	0.003	−1.735	−0.898	−0.986	−0.899	0.001
1575	4.007	0.003	11.993	5.992	4.508	6.000	0.024
1140	5.911	0.003	−2.911	−1.197	−1.324	−1.199	0.001
769	4.182	0.003	−3.182	−1.556	−1.872	−1.558	0.002
130	5.252	0.005	−0.252	−0.110	−0.111	−0.110	0.000
1642	7.132	0.005	3.868	1.448	1.340	1.452	0.003
475	3.935	0.003	4.065	2.049	1.795	2.052	0.003
973	3.737	0.008	5.263	2.722	2.301	2.734	0.016
569	4.382	0.006	−3.382	−1.616	−1.952	−1.620	0.004
1180	4.603	0.003	−3.603	−1.679	−2.038	−1.682	0.002
97	3.567	0.004	3.433	1.818	1.604	1.822	0.003

adjust for overdispersion, one needs to add the scale parameter by changing "scale = none" to "scale = factor".

What could be another reason for poor fit besides overdispersion?

- Omission of explanatory variable;
- Heterogeneity.

Solutions

- Consider different methods;
- Collapse over levels of explanatory variables;
- Transform the variables.

In our example the scale factor is 3.830. A correction through scaling gave the following results in Table 8.5.

The ratio ($\frac{value}{df}$ = 0.957 and 1.244) is much closer to a value of one. The revised parameter estimates are given in Table 8.6.

The adjusted model (overdispersed) gave the same estimates as in the unadjusted model but with a larger standard error. Hence, size is still significant in the model, but age is no longer significant.

8.5.1 Statistical Analysis of Data SAS Program

```
*** Poisson Regression;
proc genmod data=newbreast;
class hormon;
```

TABLE 8.5

Goodness of Fit[a]

	Value	df	Value/df
Deviance	652.935	682	.957
Scaled deviance	652.935	682	
Pearson chi-square	848.432	682	1.244
Scaled Pearson chi-square	848.432	682	
Log likelihood[b]	-590.028		
Akaike's Information Criterion (AIC)	1188.057		
Finite Sample Corrected AIC (AICC)	1188.115		
Bayesian Information Criterion (BIC)	1206.180		
Consistent AIC (CAIC)	1210.180		

Dependent Variable: nodes
Model: (Intercept), age, size, hormone
[a] *Information criteria are in smaller-is-better form.*
[b] *The full log likelihood function is displayed and used in computing information criteria.*

TABLE 8.6

Parameter Estimates (Overdispersed Model)

Parameter	B	Std. Error	95% Wald Confidence Interval		Hypothesis Test		
			Lower	Upper	Wald Chi-Square	df	Sig.
(Intercept)	.713	.2034	.314	1.111	12.281	1	.000
age	.005	.0036	−.002	.012	2.222	1	.136
size	.019	.0018	.016	.023	111.925	1	.000
hormon	.026	.0729	−.117	.169	.126	1	.723
(Scale)	1[a]						

Dependent variable: nodes; Model: (Intercept), age, size, hormon.
[a] *Fixed at the displayed value.*

```
model nodes=size age hormon/dist=poisson;
run;
The SAS output is presented.
```

The GENMOD Procedure

Model Information	
Dataset	WORK.NEWBREAST
Distribution	Poisson
Link Function	Log
Dependent Variable	nodes Nodes
Number of Observations Read	686
Number of Observations Used	686

The random component (distribution of responses) is Poisson. The link function is log. There are 686 items in this dataset.

Class Level Information		
Class	Levels	Values
hormon	2	had tamoxifen no tamoxifen

Criteria For Assessing Goodness Of Fit			
Criterion	DF	Value	Value/DF
Deviance	682	2611.7392	3.8295
Scaled Deviance	682	2611.7392	3.8295
Pearson Chi-Square	682	3393.7279	4.9761
Scaled Pearson X2	682	3393.7279	4.9761

Log Likelihood	2289.1118
Full Log Likelihood	−2360.1135
AIC (smaller is better)	4728.2270
AICC (smaller is better)	4728.2857
BIC (smaller is better)	4746.3505
Algorithm converged.	

There are several test statistics for the goodness of fit test. In particular the scaled deviance and the scaled Pearson are used. The fact that these values are 3.8295 and 4.9761 (different from the value of one) suggests that there is overdispersion in these data.

Analysis Of Maximum Likelihood Parameter Estimates

Parameter		DF	Estimate	Standard Error	Wald 95% Confidence Limits		Wald Chi-Square	Pr > ChiSq
Intercept		1	0.7127	0.1017	0.5134	0.9120	49.12	<.0001
size		1	0.0192	0.0009	0.0174	0.0210	447.70	<.0001
age		1	0.0053	0.0018	0.0018	0.0088	8.89	0.0029
hormon	had tamoxifen	1	0.0259	0.0365	−0.0456	0.0973	0.50	0.4782
hormon	no tamoxifen	0	0.0000	0.0000	0.0000	0.0000	.	.
Scale		0	1.0000	0.0000	1.0000	1.0000		

Note: The scale parameter was held fixed.

The fitted Poisson model is

$$log[nodes] = 0.71 + 0.019size + 0.005age + 0.026hormon$$

Size (p <.0001) and age (p = 0.0029) are significant covariates. The over-dispersion needs to be addressed. One can refit with the scale of about 4.

8.5.2 Statistical Analysis of Data R Program

```
## load the R package "readxl
library(readxl)
germanbreast = read_excel(("C:/Users/angel/Desktop/newbreast.xls"))
attach(germanbreast)
## Call "glm" for Poisson trgression
model=glm(nodes~size+age+hormon,family=poisson(link=log)) ##Poisson
Regression
```

```
summary(model)
##
## glm(formula = nodes ~ size + age + hormon, family = poisson(link =
log))
## Deviance Residuals:
##     Min          1Q         Median      3Q          Max
##     -4.1411     -1.7172     -0.7748     0.6926      9.6848
## Coefficients:
##                           Estimate    Std. Error   z value   Pr(>|z|)
## (Intercept)              0.7385771    0.1111471    6.645     3.03e-11  ***
## size                     0.0191782    0.0009064    21.159    < 2e-16   ***
## age                      0.0053035    0.0017791    2.981     0.00287   **
## hormonno tamoxifen      -0.0258570    0.0364593   -0.709     0.47820
```

The fitted Poisson model is

$$log[nodes] = 0.71 + 0.019 size + 0.005 age + 0.026 hormon$$

Size (p <.0001) and age (p = 0.0029) are significant covariates.

```
## (Dispersion parameter for Poisson family taken to be 1)
##     Null deviance: 2986.7 on 685 degrees of freedom
## Residual deviance: 2611.7 on 682 degrees of freedom
## AIC: 4728.2
## Number of Fisher Scoring iterations: 5
```

There are several statistics for the goodness of fit test. In particular, the residual deviance has a value of 2611.7 with degrees of freedom of 682. The scaled deviance 3.8295 = 2611.7/682 (different from the value of one) suggests that there is overdispersion in these data. One may want to refit this with a scaling = 2611.7/682.

8.6 Summary and Discussions

There was a time in statistical modeling when the analysis of data relied to a great extent on the normal distribution assumption when modeling the mean of continuous responses. In cases when the normal distribution seems to be a poor choice, one changed the scale to transform the outcomes to comply with the normality assumption. However, the advent of the generalized linear model negates the need for relying on the normal distribution or its transformed values to be normally distributed. The generalized linear model allows one to fit both the mean of continuous and the mean of categorical responses. It allows one to model the mean of outcomes that are not normally distributed as long as it is a member of the exponential family.

In generalized linear models and, in particular, the Poisson regression model, the mean and the variance are related. It is essential in fitting a generalized linear model that a mean-variance relation exists. If this assumed relation is violated, one can fit an overdispersed model by using the scale factor that signifies violation. Those and other correlated models are addressed in Chapter 11.

Question: What is the consequence of overdispersion in the model?

Answer: The underestimation of the variance and, hence, the standard error of the coefficient, can lead to a conclusion of significance when that may not really be the case.

8.7 Exercises

Using the data examined in this chapter, answer these questions.

1. Fit a Poisson regression model to mean nodes with covariates, grade, menopause, and tamoxifen.
2. Did the model fit?
3. Were the covariates significant?
4. Why do you think the model did or did not fit?

References

Agresti, A.: *An Introduction to Categorical Data Analysis*, 2nd ed. Wiley, New York (2007)

Agresti, A.: *Categorical Data Analysis*, 3rd ed. Wiley, New York (2013)

Anderson, D., et al.: A practitioner's guide to generalized linear models. https://www.casact.org/pubs/dpp/dpp04/04dpp1.pdf (2007)

Burnham, K.P., Anderson, D.R.: *Model Selection and Inference: A Practical Information Theoretic Approach*. Springer, New York (1998)

Gary, D.C.: *Generalized Linear Models. Predictive Modeling Applications in Actuarial Science*. Cambridge University Press, New York (2014)

Harrell, Jr. F.E.: *Regression Modeling Strategies with Applications to Linear Models, Logistic Regression, and Survival Analysis*. Springer, New York (2001)

McCullagh, P., Nelder, J.A.: *Generalized Linear Models*, 2nd ed. Chapman & Hall, London (1989)

Nelder J.A., Wedderburn, R.W.M.: Generalized linear models. *Journal of the Royal Statistical Society: Series A*, 135(3), 370–384 (1972). https://doi.org/10.2307/2344614

Sauerbrei, W., Royston, P.: Building multivariable prognostic and diagnostic models: Transformation of the predictors by using fractional polynomials *Journal of the Royal Statistical Society: Series A*, 162(1), 71–94 (1999). https://doi.org/10.1111/1467-985X.00122

Schmoor, C., Olschhewski, M., Schumacher, M. Randomized and Nonrandomized patients in clinical trials experiences with comprehensive cohort studies. *Statistics in Medicine*, 15, 263–271 (1996)

9

Modeling Repeated Continuous Observations Using GEE

9.1 Research Interest/Question

This chapter examines modeling responses measured on a continuum, when a sampling unit/individual is repeatedly measured. This considers modeling correlated observations (i.e., within-subject correlation). Correlated observations are collected in many areas of research and as such require a model that pays close attention to the extra variation or the overdispersion caused by the correlation. In addition, the different types of within-subject correlation due to repeated measurement over time on the same subject are modeled. The simplest example of such a situation is to measure the scores on a continuum before and after the study for participants who have been exposed to some process (i.e., pre- and post-design). This is referred to as the well-known two-sample paired t-test.

However, this chapter addresses processes with more than two measurements on a unit/individual. In particular, an examination of the analysis of correlated continuous responses using generalized estimating equations is demonstrated. This method of analysis is outlined through its applications to the Medicare data to address rehospitalization.

9.2 Public Health Data

The Medicare data were extracted from the Arizona State Inpatient Databases and pertain to patients admitted to a hospital for a period of four visits. These data contain information on patients discharged from Arizona hospitals between 2003 and 2005. The variable of interest is the length of stay (LOS) for patients readmitted (PNUM_R) for the same condition, for which they were initially hospitalized. Rehospitalization is relevant, as Medicare will pay for all subsequent visits for patients, except in cases where readmission occurs within one day for the same procedure. Covariates and predictors are identified based on the findings of Jencks, Williams, and Coleman

DOI: 10.1201/9781003315674-9

(2009). In particular, these data include the total number of diagnoses (NDX) and the total number of procedures performed (NPR) for each patient. Also, information on coronary atherosclerosis (DX101) and osteoarthritis (DX203) are included. An additional column (Time) indicates the times of the visits for each participant. These data are used for demonstration purposes since the outcomes are correlated. A subset of these data is shown in Table 9.1.

9.3 Two-Measurements on An Experimental Unit

Consider an experiment or a process where the sampling units are measured before and after a certain event has occurred. While the sampling units are independent, the observational measures "before" and "after" for each sampling unit are correlated (dependent). It is a repeated measures design but, since the measurements occurred twice, this design is known as the two-sample paired t-test. In this case, one is interested in the mean difference between "before" and "after" the event has occurred. The design is greatly efficient, as one could think that the difference is due to the event because it is the same unit measured twice. One does not have to question whether a sampling unit differs from one result to the next (as long as the exposure to the event before doesn't alter the identity of the sampling unit), as it is the same unit. Alternatively, we can ignore whether there are characteristics of the sampling unit that may have given rise to the result other than the event but that requires a covariate to be in the model. However, this design suggests that the change can only be attributed to the event (for example an intervention).

TABLE 9.1

A Subset of Medicare Data

PNUM_R	NDX	NPR	LOS	DX101	OSTEOARTHRITIS	Time
127	9	6	6	1	0	1
127	6	4	1	1	0	2
127	9	5	3	1	0	3
560	9	3	8	0	0	1
560	9	1	17	0	0	2
560	7	1	6	0	0	3
746	6	4	12	0	0	1
746	6	1	1	0	0	2
746	9	1	2	0	0	3
750	9	3	6	0	0	1
750	7	3	4	0	0	2
750	9	2	4	0	0	3
1117	9	6	5	1	0	1
1117	9	3	1	0	0	2
1117	9	6	4	1	0	3

For this design, the null hypothesis is written as $H_0 : \mu_{before} = \mu_{after}$, and the alternative hypothesis is that the before and after means are different from each other. The alternative can also be that one mean is less than or greater than the other. A test statistic based on a standardized difference between "before" and "after" is used to test this hypothesis. The test statistic follows a t-distribution with n-1 degrees of freedom, where *n* is the number of sampling units. The value of the statistic can be used to obtain the p-value and give some probability measure of significant difference.

9.4 Generalized Estimating Equations Models

Generalized linear models (GLMs) are applicable when the observations are independent. GLM is a framework for independent observations when the responses come from a distribution that is a member of the exponential family. A general framework for relating response and predictor variables is outlined in McCullagh and Nelder (1989) and Dobson (1983).

However, studies such as prospective cohort studies, where individuals are followed for a certain time period and the outcomes are recorded repeatedly, produce correlated observations. Such data cannot be analyzed with generalized linear models, as these outcomes are not independent Dobson (2002). There is an inherent correlation.

When taking repeated measurements, an individual *i* has an outcome at time *t* as (y_{it}) is expected to have a higher probability of having a similar outcome (y_{is}) at time *s* than outcomes from y_{it} versus y_{jt}, from different individuals *i* and *j*. The complexity brought on by correlated outcomes is often considered a challenge to making sound statistical inferences and as such cannot be ignored. This chapter concentrates on the use of the *generalized estimating equation* to model continuous correlated observations using a variety of correlation structures. There are two basic approaches for modeling correlated observations. The first approach is GEE and is presented in this chapter. The second approach is random-effects model and is discussed in Chapters 10 and 11.

One method of modeling correlated observations is to use GEE estimates to obtain the regression coefficients through the random component as in a generalized linear model. The GEE relies on a mean variance relation versus GLM, which relies on a distribution from the exponential family. This approach gives rise to marginal models or population-averaged models for correlated data. These models concentrate on modeling the mean just as in generalized linear models.

The second approach, random-effects model, addresses the correlation through the systematic component. Such approach gives rise to the subject-specific model. These models are considered modeling the conditional mean of some subpopulation (see Ballinger 2004).

9.4.1 Generalized Estimating Equations and Covariance Structure

The generalized estimating equations (GEEs) approach by Liang and Zeger (1986) is a method for analyzing correlated outcomes through the random component of the model. It relies on an appropriate working correlation matrix to account for the within-subject correlations. While the standard linear model uses a set of equations based on the maximum likelihood estimation to obtain the regression coefficients, the GEE model uses a set of equations based on a quasi-likelihood (almost likelihood) approach. This approach relies on presenting a working correlation matrix among the outcomes per subject. It iteratively solves a system of equations based on quasi-likelihood distributional assumptions. It is driven through the use of a working correlation matrix as developed in Zeger and Liang (1992).

9.4.2 Working Correlation Matrices

The GEE estimates rely on the use of a matrix referred to as the *working correlation matrix*. It represents a structure of the repeated measures on a subject. Consider three sampling units where each sampling unit is measured four times, presenting a correlation matrix of dimension four. Each sampling unit i has a correlation matrix of the form, as $i = 1, 2, 3$;

$$\Sigma_{i**} = \begin{bmatrix} \rho_{11} & \rho_{12} & \rho_{13} & \rho_{14} \\ \rho_{21} & \rho_{22} & \rho_{23} & \rho_{24} \\ \rho_{31} & \rho_{32} & \rho_{33} & \rho_{34} \\ \rho_{41} & \rho_{42} & \rho_{43} & \rho_{44} \end{bmatrix}$$

But the overall covariance matrix (for all units) is a block diagonal matrix of dimension 12 by 12:

$$\Sigma_{***} = \begin{bmatrix} \Sigma_{1**} & 0 & 0 \\ 0 & \Sigma_{2**} & 0 \\ 0 & 0 & \Sigma_{3**} \end{bmatrix}$$

The off diagonal block is zero as they represent the correlation among different sampling units. The matrix Σ_{i**} includes the assumption made about the association between the observations on a subject. The model assumes that the form of the relation, and not necessarily the degree of the relation, is the same for all subjects. Therefore, if one assumes that there is compound symmetry for one subject, one assumes this holds for all subjects. In reality, however, even if the structure is the same, the strength of that correlation may differ across subjects. Thus, the model takes the average across all subjects and uses that as the correlation matrix for the data. The typical working correlations for GEE models are independence, compound symmetry or exchangeability, autoregressive AR (1), unstructured, and user-defined correlation structure defined as follows:

- *Independence structure* indicates that repeated observations are uncorrelated. It is the simplest form of working correlation, namely, the identity matrix of dimension T (T is the maximum number of the repeated measures). This form indicates that the longitudinal data are not correlated.

$$\Sigma_{i^{**}} = \begin{bmatrix} 1 & 0 & 0 & 0 \\ 0 & 1 & 0 & 0 \\ 0 & 0 & 1 & 0 \\ 0 & 0 & 0 & 1 \end{bmatrix}$$

In general, this is not very intuitive and would be difficult to accept for repeated measures on the same subject.

- *Exchangeable* indicates that the correlation between any two outcomes of the ith subject/unit is the same. So, if a subject is measured four times, then the correlation (ρ) of observations at times (1, 2); (1, 3); (1, 4); (2, 3); (2, 4); and (3, 4) are all the same, resulting in

$$\Sigma_{i^{**}} = \begin{bmatrix} 1 & \rho & \rho & \rho \\ \rho & 1 & \rho & \rho \\ \rho & \rho & 1 & \rho \\ \rho & \rho & \rho & 1 \end{bmatrix}$$

- *Autoregressive*, or *AR (1)*, indicates that the correlation between any two observations is assumed to be less as they become further apart, which is measured by $\rho^{\{\text{time period difference}\}}$. Since it depends only on one parameter, ρ, it is a very parsimonious representation for longitudinal data;

$$\Sigma_{i^{**}} = \begin{bmatrix} 1 & \rho & \rho^2 & \rho^3 \\ \rho & 1 & \rho & \rho^2 \\ \rho^2 & \rho & 1 & \rho \\ \rho^3 & \rho^2 & \rho & 1 \end{bmatrix}$$

This autoregressive assumption (also referred to as a "transitional model") is used when the analysis must account for a time dependency, as discussed in Hardin and Hilbe (2003), and Pan and Connett (2002);

- *Unstructured*, or unspecified, indicates that the correlation within any two responses is unknown and must be estimated. Thus, no structural form is assumed, and it may be that all correlations per subject/unit are different.

$$\Sigma_{i**} = \begin{bmatrix} 1 & \rho_{12} & \rho_{13} & \rho_{14} \\ \rho_{21} & 1 & \rho_{23} & \rho_{24} \\ \rho_{31} & \rho_{32} & 1 & \rho_{34} \\ \rho_{41} & \rho_{42} & \rho_{43} & 1 \end{bmatrix}$$

This unstructured form is the most efficient, but is only useful if the numbers of time points are small; otherwise, there are too many parameters to be estimated;

- *User-defined* structure indicates that the data analyst decides the working correlation matrix, maybe from past studies or from experience. This choice of user-defined working correlation matrix is not well-advised, since it can easily lead to non-convergence.

In summary, modeling the correlation with the GEE approach accounts for the association across time and the association between observations for the same subject. As such, it allows an arbitrary working correlation structure for the correlation matrix of a subject's outcomes. We make use of an average of these correlation matrices.

9.5 Statistical Data Analysis in SAS and R

The GEE method of obtaining regression estimates to fit marginal models for the length of stay (LOS) during hospitalization and rehospitalization within 30 days is demonstrated. The resulting models account for the correlation inherent in repeated observations of an individual through the use of different working correlation matrices. These common working correlation matrices, compound symmetry, autoregressive of order one, and unstructured, are used in these statistical programs. There are two statistical programs presented: SAS and R. We fit

$$y_i = \beta_0 + \beta_1 X_{i1} + \dots \beta_p X_{ip} + \varepsilon_i$$

$$v\left(\varepsilon_i\varepsilon_j\right) = f\left(\mu\right)$$

9.5.1 Statistical Analysis of Data SAS Program

The GEE Model with three types of working correlation matrices using:

1. compound symmetry in the first GEE model, (same correlation between all observations);
2. AR (1) in the second GEE model, (declining correlation between observations based on distance in time); and
3. Unstructured correlation (UN) in the third GEE model is reported (no definite pattern).

GEE Model 1 with CS:

```
proc genmod data=anhdata;
class DX101(ref="0") time(ref="1") DX203(ref="0") PNUM_R;
model LOS=NDX NPR DX101 DX203 time NDX*time NPR*time;
repeated subject=PNUM_R/within =time type=cs corrw;
run;
```

We fit compound symmetry for modeling length of stay with covariates NDX and NPR and predictors DX101, DX203, and time, with interaction NDX and time, as well as NPR and time. PNUM_R is the identification for each patient.

The GENMOD Procedure

Model Information	
Dataset	WORK.ANHDATA
Distribution	Normal
Link function	Identity
Dependent variable	LOS
Number of observations read	4875
Number of observations used	4875

The distribution for the response length of stay at a particular time point is assumed to be normal, with identity link; and 4875 observations for 1625 patients at 3 repeated measures times.

Class Level Information		
Class	Levels	Values
DX101	2	1 0
times	3	2 3 1
DX203	2	1 0
PNUM_R	1625	127 560 746 750 1117 1395 1568 2076 2390 2413 3008 3123 3710 3970 11274 11279 11787 13420 13436 13761 14955 16160 16464 16971 17748 18638 18697 19349 19674 19730 20112 20973 21410 21800 ...

The DX101, times, and DX203 are categorical variables, and PNUM_R is the identification number. The number of levels for each variable is defined.

GEE Model Information	
Correlation structure	Exchangeable
Within-subject effect	times (3 levels)
Subject effect	PNUM_R (1625 levels)
Number of clusters	1625
Correlation matrix dimension	3
Maximum cluster size	3
Minimum cluster size	3

Exchangeable is another name for compound symmetry. It allows a covariate of each observation, which is not independent, to have the same correlation. The matrix serves to bring these observations together from each experimental unit. Since each patient provides three observations, we have a correlation matrix of dimension 3. All the patients had the same number of observations so the maximum = minimum = 3. PNUM_R is cluster (patient) identity.

Algorithm converged.		
Working Correlation Matrix		
	Col1	Col2	Col3
Row1	1.0000	0.1812	0.1812
Row2	0.1812	1.0000	0.1812
Row3	0.1812	0.1812	1.0000

Exchangeable Working Correlation	
Correlation	0.1811822741

The compound symmetry assumes a common correlation. The estimated value is 0.18127. Irimata and Wilson (2018) suggest that such a correlation size is too large to be ignored.

Analysis Of GEE Parameter Estimates							
Empirical Standard Error Estimates							
Parameter		Estimate	Standard Error	95% Confidence Limits		Z	Pr > |Z|
Intercept		−2.3816	0.4916	−3.3452	−1.4180	−4.84	<.0001
NDX		0.7374	0.0505	0.6385	0.8363	14.62	<.0001
NPR		1.0000	0.1102	0.7840	1.2159	9.08	<.0001
DX101	1	−3.2601	0.2390	−3.7285	−2.7917	−13.64	<.0001
DX101	0	0.0000	0.0000	0.0000	0.0000	.	.
DX203	1	0.8764	0.2132	0.4586	1.2942	4.11	<.0001
DX203	0	0.0000	0.0000	0.0000	0.0000	.	.
time	2	−0.5404	0.6442	−1.8030	0.7221	−0.84	0.4015
time	3	−1.7200	0.7350	−3.1606	−0.2794	−2.34	0.0193
time	1	0.0000	0.0000	0.0000	0.0000	.	.
NDX*time	2	0.1215	0.0704	−0.0164	0.2594	1.73	0.0842
NDX*time	3	0.2487	0.0801	0.0917	0.4057	3.10	0.0019
NDX*time	1	0.0000	0.0000	0.0000	0.0000	.	.
NPR*time	2	0.0279	0.1387	−0.2440	0.2997	0.20	0.8409
NPR*time	3	0.2054	0.1541	−0.0967	0.5075	1.33	0.1826
NPR*time	1	0.0000	0.0000	0.0000	0.0000	.	.

The GEE model with compound symmetry suggests that the number of procedures, number of diagnoses, coronary atherosclerosis, and osteoarthritis have a statistically significant impact on the length of stay (see p-values). Moreover, the number of diagnoses changes over time (see interactions). So, a patient having had a coronary atherosclerosis is likely to have a shorter stay in hospital as opposed to a patient who did not. Whereas a patient having osteoarthritis is more likely to have a longer stay than a patient who did not.

GEE Model 2 with AR(1):

```
proc genmod data=anhdata;
class DX101(ref="0") time(ref="1") DX203(ref="0") PNUM_R ;
model LOS=NDX NPR DX101 DX203 time NDX*time NPR*time;
repeated subject=PNUM_R/within =time type=ar(1) corrw;
run;
```

We fit an auto regressive model of order one (AR (1)) for length of stay (LOS) with covariates NDX and NPR and predictors DX101, DX203 = osteoarthritis and time with interaction: NDX and time and NPR and time. PNUM_R is the identification for each patient.

Working Correlation Matrix			
	Col1	Col2	Col3
Row1	1.0000	0.1967	0.0387
Row2	0.1967	1.0000	0.1967
Row3	0.0387	0.1967	1.0000

Auto regressive 1 model: Since each patient provided three observations, we have a correlation matrix of dimension 3. For each patient, time 1 and time 2 have the same correlation as time 2 and time 3 = 0.1967. However, time 1 and time 3 are not adjacent but two periods apart. They have their correlation squared $=(0.1967)^2 = 0.0387$.

Analysis of GEE Parameter Estimates							
Empirical Standard Error Estimates							
Parameter		Estimate	Standard Error	95% Confidence Limits		Z	Pr > |Z|
Intercept		−2.3078	0.4921	−3.2724	−1.3433	−4.69	<.0001
NDX		0.7445	0.0500	0.6465	0.8425	14.89	<.0001
NPR		0.9662	0.1104	0.7497	1.1827	8.75	<.0001
DX101	1	−3.3473	0.2415	−3.8205	−2.8740	−13.86	<.0001
DX101	0	0.0000	0.0000	0.0000	0.0000	.	.
DX203	1	0.8442	0.2147	0.4233	1.2651	3.93	<.0001
DX203	0	0.0000	0.0000	0.0000	0.0000	.	.
time	2	−0.5027	0.6446	−1.7662	0.7608	−0.78	0.4355
time	3	−1.8005	0.7234	−3.2184	−0.3826	−2.49	0.0128
time	1	0.0000	0.0000	0.0000	0.0000	.	.
NDX*time	2	0.0975	0.0708	−0.0412	0.2363	1.38	0.1683
NDX*time	3	0.2510	0.0771	0.0999	0.4022	3.26	0.0011
NDX*time	1	0.0000	0.0000	0.0000	0.0000	.	.
NPR*time	2	0.0719	0.1378	−0.1982	0.3420	0.52	0.6018
NPR*time	3	0.2180	0.1595	−0.0947	0.5306	1.37	0.1719
NPR*time	1	0.0000	0.0000	0.0000	0.0000	.	.

The GEE model with autoregressive (1) suggests that the number of procedures, the number of diagnoses, coronary atherosclerosis, osteoarthritis, and time statistically significantly impact the length of stay (see p-values). Moreover, the number of diagnoses changes over time (see interactions).

GEE Model 3 with UN:

```
proc genmod data=anhdata;
class DX101(ref="0") time(ref="1") DX203(ref="0") PNUM_R ;
model LOS=NDX NPR DX101 DX203 time NDX*time NPR*time;
repeated subject=PNUM_R/within =time type=un corrw;
run;
```

We fit unstructured correlation for modeling length of stay with covariates NDX and NPR and predictors DX101 and DX203, and time with interaction NDX and time and NPR and time. PNUM_R is the identification for each patient.

Working Correlation Matrix

	Col1	Col2	Col3
Row1	1.0000	0.1723	0.1484
Row2	0.1723	1.0000	0.2254
Row3	0.1484	0.2254	1.0000

Unstructured correlation: Since each patient provided three observations, we have a correlation matrix of dimension 3. For each patient, the correlation between any two times has no common relation as in compound symmetry or autoregressive of order 1. We have $(3 \times 2)/2 = 3$ different correlations: 0.1723, 0.1484, and 0.2254.

Analysis Of GEE Parameter Estimates

Empirical Standard Error Estimates

Parameter	Estimate	Standard Error	95% Confidence Limits		Z	Pr > |Z|
Intercept	−2.3747	0.4903	−3.3357	−1.4137	−4.84	<.0001
NDX	0.7430	0.0501	0.6448	0.8412	14.83	<.0001

NPR		0.9838	0.1100	0.7681	1.1995	8.94	<.0001
DX101	1	−3.2467	0.2384	−3.7140	−2.7795	−13.62	<.0001
DX101	0	0.0000	0.0000	0.0000	0.0000	.	.
DX203	1	0.8755	0.2129	0.4583	1.2928	4.11	<.0001
DX203	0	0.0000	0.0000	0.0000	0.0000	.	.
time	2	−0.4405	0.6430	−1.7007	0.8197	−0.69	0.4933
time	3	−1.7200	0.7296	−3.1499	−0.2901	−2.36	0.0184
time	1	0.0000	0.0000	0.0000	0.0000	.	.
NDX*time	2	0.1038	0.0701	−0.0337	0.2412	1.48	0.1389
NDX*time	3	0.2410	0.0790	0.0862	0.3958	3.05	0.0023
NDX*time	1	0.0000	0.0000	0.0000	0.0000	.	.
PR*time	2	0.0373	0.1390	−0.2352	0.3098	0.27	0.7884
NPR*time	3	0.2245	0.1550	−0.0793	0.5284	1.45	0.1476
NPR*time	1	0.0000	0.0000	0.0000	0.0000	.	.

> The GEE model with unstructured correlation suggests that the number of procedures, the number of diagnoses, coronary atherosclerosis, osteoarthritis, and time statistically significantly impact the length of stay. (This is based on the p-values.) More so, the number of diagnoses changes as it relates to LOS over time (see interactions).

9.5.2 Statistical Analysis of Data R Program

```
## Load the library "readxl"
library(readxl)
## read the data into R
Anhdata = read_excel(("Data file path/anhdata.xls"))
attach(anhdata)
# load the GEE package
library(geepack)


## GEE Model 1 with CS
## Call "geeglm" to fit GEE model with "exchangable"
correlation
CS = geeglm(LOS~NDX+NPR+factor(DX101)+factor(DX203)
      +factor(time)+NDX*factor(time)+NPR*factor(time),
      id=factor(PNUM_R),data = anhdata,corstr=
         "exchangeable")
summary(CS)
```

> We fit compound symmetry for modeling LOS with covariates NDX and NPR and predictors DX101 and DX203, and time with interaction NDX and time as well as NPR and time.

```
## Call:
## geeglm(formula = LOS ~ NDX + NPR + factor(DX101) +
                     factor(DX203) +
## factor(time) + NDX * factor(time) + NPR * factor(time),
## data = anhdata, id = factor(PNUM_R), corstr =
                     "exchangeable")
## Coefficients:
##                       Estimate  Std.err   Wald     Pr(>|W|)
## (Intercept)           -2.38161  0.49164   23.467   1.27e-06 ***
## NDX                    0.73742  0.05045   213.638  < 2e-16  ***
## NPR                    0.99996  0.11018   82.374   < 2e-16  ***
## factor(DX101)1        -3.26011  0.23900   186.073  < 2e-16  ***
## factor(DX203)1         0.87640  0.21319   16.900   3.94e-05 ***
## factor(time)2         -0.54041  0.64417   0.704    0.40151
## factor(time)3         -1.72002  0.73501   5.476    0.01928  *
## NDX:factor(time)2      0.12149  0.07035   2.982    0.08418  .
## NDX:factor(time)3      0.24867  0.08010   9.639    0.00191  **
## NPR:factor(time)2      0.02786  0.13872   0.040    0.84085
## NPR:factor(time)3      0.20543  0.15414   1.776    0.18261
## Signif. codes: 0 '***' 0.001 '**' 0.01 '*' 0.05 '.' 0.1 ' ' 1
##
```

> The GEE model with compound symmetry suggests that the number of procedures, the number of diagnoses, coronary atherosclerosis, osteoarthritis and time statistically significantly impact the length of stay. (This is based on the p-values.) More so, the number of diagnoses changes over time (see interactions). DX101 is a binary variable, so the output has information for one of the two categories. So, a patient having coronary atherosclerosis is likely to have a shorter stay in hospital as opposed to a patient who does not. Whereas a patient with osteoarthritis is more likely to have a longer stay than a patient who does not.

```
## Estimated Scale Parameters:
##        Estimate   Std.err
## (Intercept)   38.02      4.562
## Correlation: Structure = exchangeable Link = identity
```

```
## Estimated Correlation Parameters:
##       Estimate  Std.err
## alpha            0.1812    0.02311
## Number of clusters:         1625 Maximum cluster size: 3
```

> The compound symmetry assumes a common correlation. The estimated value is 0.1812. Irimata and Wilson (2018) suggest that such a correlation size is too large to be ignored.

```
## GEE Model 2 with AR1:
## Call "geeglm" to fit GEE model with "AR1" correlation
AR1 = geeglm(LOS~NDX+NPR+factor(DX101)+factor(DX203)+fac
tor(time)+NDX*factor(time)+NPR*factor(time),id=factor(P
NUM_R),data = anhdata,corstr="ar1")
summary(AR1)

## Call:
## geeglm(formula = LOS ~ NDX + NPR + factor(DX101) +
                   factor(DX203) +
##      factor(time) + NDX * factor(time) + NPR *
              factor(time),
##      data = anhdata, id = factor(PNUM_R), corstr =
              "ar1")
##
```

> Auto regressive 1 model: Since each patient provided three observations, we have a correlation matrix of dimension 3. For each patient, time 1 and time 2 have the same correlation (ρ) as time 2 and time 3 also (ρ). However, time 1 and time 3 are not adjacent, but they are two periods apart, so the correlation is $*\rho = \rho^2$. They have their correlation squared.

```
## Coefficients:
##                  Estimate  Std.err  Wald   Pr(>|W|)
## (Intercept)      -2.3172   0.4930   22.10  2.6e-06  ***
## NDX               0.7429   0.0501  219.82  < 2e-16  ***
## NPR               0.9690   0.1104   77.06  < 2e-16  ***
## factor(DX101)1   -3.2941   0.2412  186.56  < 2e-16  ***
## factor(DX203)1    0.8625   0.2160   15.95  6.5e-05  ***
```

```
## factor(time)2        -0.4609   0.6450   0.51     0.4748
## factor(time)3        -1.7934   0.7232   6.15     0.0131    *
## NDX:factor(time)2     0.0927   0.0711   1.70     0.1922
## NDX:factor(time)3     0.2509   0.0771  10.60     0.0011    **
## NPR:factor(time)2     0.0718   0.1371   0.27     0.6004
## NPR:factor(time)3     0.2187   0.1594   1.88     0.1698
```

The GEE model with autoregressive (1) suggests that the number of procedures, the number of diagnoses, coronary atherosclerosis, osteoarthritis, and time statistically significantly impact the length of stay (see
p-values). More so, the number of diagnoses changes over time (see
interactions).

```
## Correlation: Structure = ar1  Link = identity
## Estimated Correlation Parameters:
##      Estimate    Std.err
## alpha      0.219 0.0279
## Number of clusters:  1625  Maximum cluster size: 3
```

Auto regressive 1 model: Since each patient provided three observations, we have a correlation matrix of dimension 3. For each patient,
time 1 and time 2 have the same correlation as time 2 and time 3 = 0.219.
However, time 1 and time 3 are not adjacent but are two periods apart.
They have their correlation squared $= (0.219)^2 = 0.048$. The standard
error of 0.0279 suggests that they are statistically significant.

GEE Model 3 with UN:

```
## Call "geeglm" to fit GEE model with "unstructured"
        correlation
UN<-geeglm(LOS~NDX+NPR+factor(DX101)+factor(DX203)+fac
        tor(time)
    +NDX*factor(time)+NPR*factor(time),id=factor(P
            NUM_R),
    data = anhdata,corstr="un")
summary(UN)

## Call:
## geeglm(formula = LOS ~ NDX + NPR + factor(DX101) +
                factor(DX203) +
```

```
## factor(time) + NDX * factor(time) + NPR *
                 factor(time),
## data = anhdata, id = factor(PNUM_R), corstr = "un")
```

We fit UN for modeling LOS with covariates NDX and NPR and predictors DX101 and DX203, and time with interaction: NDX and time and NPR and time. PNUM_R is the identification for each patient.

```
## Coefficients:
##                      Estimate Std.err  Wald     Pr(>|W|)
## (Intercept)          -2.3741  0.4903   23.45    1.3e-06  ***
## NDX                   0.7431  0.0501   220.08   < 2e-16  ***
## NPR                   0.9836  0.1101   79.89    < 2e-16  ***
## factor(DX101)1       -3.2491  0.2384   185.72   < 2e-16  ***
## factor(DX203)1        0.8748  0.2129   16.89    4.0e-05  ***
## factor(time)2        -0.4425  0.6430   0.47     0.4913
## factor(time)3        -1.7206  0.7295   5.56     0.0184   *
## NDX:factor(time)2     0.1040  0.0701   2.20     0.1382
## NDX:factor(time)3     0.2410  0.0790   9.32     0.0023   **
## NPR:factor(time)2     0.0374  0.1391   0.07     0.7878
## NPR:factor(time)3     0.2244  0.1551   2.10     0.1478
##
```

The GEE model with unstructured correlation suggests that the number of procedures, the number of diagnoses, coronary atherosclerosis, osteoarthritis, and time statistically significantly impact the length of stay. (This is based on the p-values.) More so, the number of diagnoses changes as it relates to LOS over time (see interactions).

```
## Signif. codes: 0 '***' 0.001 '**' 0.01 '*' 0.05 '.'
0.1 ' ' 1
##      Estimated Scale Parameters:
##       Estimate  Std.err
## (Intercept)  38       4.56
## Correlation: Structure = unstructured  Link =
                 identity
## Estimated Correlation Parameters:
##       Estimate Std.err
## alpha.1:2    0.171     0.0335
## alpha.1:3    0.148     0.0276
## alpha.2:3    0.224     0.0464
## Number of clusters: 1625  Maximum cluster size: 3
```

> **Unstructured correlation:** Since each patient provided three observations, we have a correlation matrix of dimension 3. Each patient, the correlation between any two times has no common relation as in compound symmetry or autoregressive of order 1. We have $(3 \times 2)/2 = 3$ different correlations: 0.171, 0.148, and 0.224. Their standard errors suggest that they are statistically significant correlations.
>
> The GEE model with unstructured correlation suggests that the number of procedures, the number of diagnoses, coronary atherosclerosis, osteoarthritis, and time statistically significantly impact the length of stay. (This is based on the p-values.) Moreover, the number of diagnoses changes as it relates to LOS over time (see interaction).

9.6 Research/Questions and Comments

WE observe that the length of stay at each hospitalization of a patient is related to their number of diagnoses and the number of procedures. These observations are correlated. In analyzing these data, we demonstrate the fit of GEE models to *Medicare* data using SAS and R. The models are fitted using generalized estimating equations through the random component. When accounting for the dependency in these data, the models treat the correlation as a nuisance, and no direct impact is addressed. In particular, the correlation is addressed by assuming that a specific pattern exists among the observations. As the subjects are repeatedly measured, one expects an inherent correlation among the responses per subject.

The GEE regression models are fitted to the rehospitalization data using three different (exchangeability, autoregressive, and unstructured) working correlation matrices. The results with predictors, NDX, NPR, DX203, and DX101 are similar. A summary of the results is given in Table 9.2. From Table 9.2, NDX and NPR are significant covariates in three models where the data determine the estimates of the correlation. As the GEE regression model is a population-averaged or marginal model, the parameter estimates describe differences in the mean response between subpopulations that differ according to the values of the predictors.

These parameter estimates are not constructed to make conclusions about changing values for individuals. It gives an idea about the average. These GEE models assume that the covariates are time-independent. The models did not take into account the fact that these covariates are time-dependent.

TABLE 9.2

Estimates and Standard Errors for GEE Regression Models

	Exchangeability		Autoregressive (1)		Unstructured	
	Estimate	SE	Estimate	SE	Estimate	SE
(Intercept)	−6.485	.6395	−7.111	.6735	−7.228	.6501
[time=1]	1.720	.7350	1.972	.7431	1.800	.7081
[time=2]	1.180	.7079	1.222	.7603	1.115	.6912
[time=3]	0[a]	.	0[a]	.	0[a]	.
[DX101=0]	3.260	.2390	3.709	.2433	3.806	.2338
[DX101=1]	0[a]	.	0[a]	.	0[a]	.
[DX203=0]	−.876	.2132	−.720	.2090	−.706	.2045
[DX203=1]	0[a]	.	0[a]	.	0[a]	.
NDX	.986	.0602	1.008	.0634	1.001	.0596
NPR	1.205	.1340	1.171	.1423	1.206	.1353
[time=1] * NDX	−.249	.0801	−.258	.0791	−.252	.0755
[time=2] * NDX	−.127	.0786	−.130	.0821	−.120	.0756
[time=3] * NDX	0[a]	.	0[a]	.	0[a]	.
[time=1] * NPR	−.205	.1541	−.245	.1666	−.215	.1557
[time=2] * NPR	−.178	.1646	−.204	.1865	−.181	.1646
[time=3] * NPR	0[a]	.	0[a]	.	0[a]	.
(Scale)	38.106		38.057		38.056	

9.7 Exercises

Use the Medicare data to fit the following:

1. Use the length of stay as the response with NDX, NPR, DX101, and DX203, with time as a factor to:
   a. Fit a standard logistic regression model.
   b. Fit a GEE model with independence structure.
   c. Fit a GEE model with exchangeability structure.
   d. Fit a GEE model with an unstructured working correlation matrix.
   e. Fit a GEE model with an autoregressive (1) correlation matrix.
2. How do these parameter estimates compare?
3. What patterns did you observe?

References

Ballinger G.A.: Using generalized estimating equations for longitudinal data analysis. *Organ Res Methods*, 7, 127–150 (2004)

Dobson, A.J.: *An Introduction to Statistical Modelling*. Chapman & hall, London (1983).

Dobson, A.J.: *An Introduction to Generalized Linear Models*. Chapman & Hall/CRC, Boca Raton (2002)

Hardin, J.W., Hilbe, J.M.: *Generalized Estimating Equations*. Wiley, New York (2003)

Irimata, K.M., Wilson, J. (2018). Identifying intraclass correlations necessitating hierarchical modeling. *Journal of Applied Statistics*, 45(4), 626–641.

Jencks, S.F., Williams, M.V., Coleman, E.A.: Rehospitalizations among patients in the Medicare fee-for-service program [published correction appears in *N Engl J Med*. 2011 Apr 21;364(16):1582]. *The New England Journal of Medicine*, 360(14), 1418–1428 (2009)

Liang, K.Y., Zeger, S.L.: Longitudinal data analysis using generalized linear models. *Biometrika*, 73(1), 13–22 (1986)

McCullagh, P., Nelder, J.: *Generalized Linear Models*. 2nd ed. Chapman and Hall, London (1989)

Pan, W., Connett, J.E.: Selecting the working correlation structure in generalized estimating equations with application to the lung health study. *Statistica Sinica*, 12(2), 475–490 (2002)

Zeger, S.L., Liang, K.Y.: An overview of methods for the analysis of longitudinal data. *Statistics in Medicine*, 11(14–15), 1825–1839 (1992)

10

Modeling for Correlated Continuous Responses with Random-Effects

---

10.1 Research Interest/Question

In this chapter, correlated continuous observations are addressed with the use of random-effects modeling. A linear mixed-effects (LME) model is presented to analyze correlated continuous responses that may have arisen as a result of clustering or through a hierarchical structure of the data or from longitudinal studies. In all instances, the observations violate the independence assumption, and the classical regression model with the use of least-squares is no longer valid for analyzing such data.

Correlated responses require a method that incorporates the inter-dependence, often referred to as the intra-class correlation (ICC), within units. In this chapter, such intraclass correlation is addressed through the use of random-effects, or mixed-effects, models.

The **Hospital Doctor Patient** (HDP) data are revisited as the use of mixed-effects models is demonstrated. These data are derived from a multilevel data structure, where the patients are nested within the doctors and the doctors are nested within hospitals. These data are used in Chapters 4, 5, and 6 to illustrate the t-test, analysis of variance (ANOVA), analysis of covariance (ANCOVA), and linear regression model without giving consideration to the hierarchical structure of the data. In this chapter, the focus is on a linear model addressing the relation between a response variable in tumor size (i.e., "tumorsize") and a covariate and the length of hospital stay (i.e., "LengthofStay"), while adjusting for other covariates and addressing the nesting structure present due to patients within doctors.

10.2 Hospital Doctor Patient Data

10.2.1 Data Source and Previous Analyses

A detailed description of the HDP data is available at https://stats.idre .ucla.edu/r/codefragments/mesimulation/#tumor. These simulated data are based on a three-level hierarchical structure with 8525 patients nested within 407 doctors, and the doctors are nested within 35 hospitals. The data are used as examples of multiple nesting structures: at the 2-level data structure and at the 3-level data structure. The outcome random variable, tumor size in millimeters (mm), is a measure of disease severity and follows a Gaussian distribution. The level-1 predictors are patients' gender, family history (FamilyHx), length of hospital stay (LengthofStay), smoking history (SmokingHx), and cancer stage (CancerStage). These patients (PID) are nested with the doctors (DID) at level-2. The data structure with ten patients is given in Table 10.1. There are ten patients nested within the first doctor. Some of the characteristics of these patients are also given in Table 10.1. HID represent the hospital ID and DID represents doctor ID.

In Chapter 4, an analysis of these HDP data examined the difference in tumor size on gender while ignoring the hierarchical structure. The average tumor size is 72.08 mm for females (n = 5115) and 69.08 mm for males (n = 3410). As such, one may conclude that the tumor size is statistically significantly different (difference of 3.01; p-value <0.00) based on gender.

Recall in Chapter 5, the HPD data revealed that tumor size differs across smoking history ("current", "former", and "never"). The average tumor size is 83.77 mm (SE = .242) for current smoker, 72.02 mm (SE = .243) for former smoker, and 66.20 mm (SE = .139) for those who never smoked. The model revealed significant differences in tumor sizes among the different kinds of smokers (p-value <0.001). This follows from the F-test in the one-way ANOVA in Chapter 5. Additionally, the interaction between cancer (four cancer stages) and "smoking history" using a two-way ANOVA was checked and found not to be statistically significant at the usual type I error level of 0.10 for tests of interaction (p-value = 0.1004; F = 1.77). As such, it implies that the difference in smoking history is roughly consistent across all stages of cancer. However, there are statistically significant differences in tumor sizes

TABLE 10.1

A Subset of the HPD Data for Illustration

Tumor size (mm)	FamilyHx	SmokingHx	Gender	Cancer Stage	Length of Stay (days)	DID	HID
67.981	no	former	male	II	6	1	1
64.702	no	former	female	II	6	1	1
51.567	no	never	female	II	5	1	1
86.438	no	former	male	I	5	1	1
53.400	no	never	male	II	6	1	1
51.657	no	never	male	I	5	1	1
78.917	no	current	female	II	4	1	1
69.833	no	former	male	II	5	1	1
62.853	yes	former	male	II	6	1	1
71.778	no	never	male	II	7	1	1

among smoking history and among the stages of cancer. In this analysis, the hierarchical structure of the data was ignored in the analysis of the data.

10.2.2 Nesting Structure

There are 8525 patients nested within 407 doctors. The 407 doctors are nested within 35 hospitals, as illustrated in Figure 10.1.

The nesting structure considered is a three-level hierarchical design. The data structure in Figure 10.1 is given in Table 10.2. The number of doctors is shown in all 35 hospitals. From this table, there are ten doctors for each of the hospitals 1 and 2, with12 and 15, for hospitals 3 and 4, etc.

A further illustration of the multilevel data structure in HDP data, Table 10.3, displays the first 454 patients nested within the first 20 doctors who are nested in the first two hospitals.

Consider tumor size as the response variable of interest and the length of hospital stay as a predictor variable in demonstrating a fit of a linear mixed-effects model. This two-dimensional relationship is depicted in Figure 10.2 for the 230 patients who are nested within the first nine doctors. At a glance of Figure 10.2, one can see the intercepts and slopes from these nine doctors are different from each other. If the information on the 407 doctors is considered together with the fixed-effects in hospital stay and a random-intercept in doctors, then it represents one of the simplest of the mixed-effects models. However, we consider a model that presents an overall relationship between the tumor size and the length of stay while adjusting for age, FamilyHx,

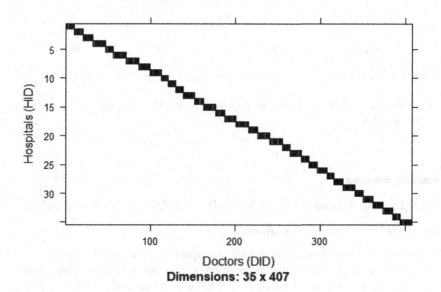

FIGURE 10.1
Nesting Structure for 407 Doctors Nested within 35 Hospitals

TABLE 10.2

Nesting Structure of 407 Doctors within 35 Hospitals

Hospital ID (HID)	# Doctors	Hospital ID (HID)	# Doctors	Hospital ID (HID)	# Doctors
1	10	13	13	25	10
2	10	14	11	26	11
3	12	15	14	27	8
4	15	16	11	28	11
5	9	17	13	29	14
6	15	18	15	30	10
7	15	19	11	31	11
8	13	20	14	32	12
9	13	21	15	33	11
10	8	22	9	34	9
11	9	23	13	35	14
12	9	24	9		

TABLE 10.3

A Subset of the Multilevel Data Structure

Hospital ID (HID)	Doctor ID (DID)	#pts	Hospital ID (HID)	Doctor ID (DID)	#pts	Hospital ID (HID)	Doctor ID (DID)	#pts
1	1	28	1	5	18	1	9	23
1	2	32	1	6	34	1	10	22
1	3	6	1	7	27	2	11	32
1	4	30	1	8	32	2	12	2
2	13	20	2	15	29	2	17	35
2	14	30	2	16	20	2	18	19
2	19	11	2	20	4			

SmokingHx, and stage of cancer across doctors. It represents a linear-mixed effects model.

10.3 Linear Mixed-Effects Models and Parameter Estimation

10.3.1 Intra-Cluster Correlation

One purpose of multilevel modeling is to quantify the correlation as it exists at the different levels of the hierarchical structure, which is referred to as the intra-class correlation (ICC). The ICC is a measure of correlation among individuals' outcome measures within the cluster or within the nested structure. For example, ICC measures the strength of the relationship among

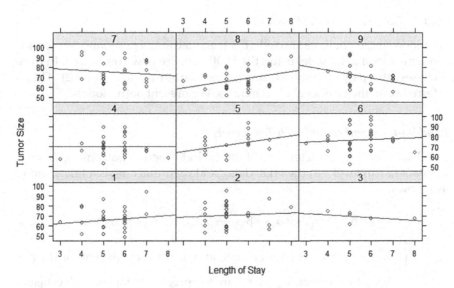

FIGURE 10.2
Tumor Sizes v. Length of Hospital Stay (Days) for Patients Nested within the First Nine Doctors

patients nested within doctors in the HDP data. The ICC (denoted by) ρ_{ICC} is an important measure in nested data analysis. It is defined as the proportion of variation among outcomes between-cluster variance (denoted by τ^2) over the total variation (i.e., the sum of the between-cluster variance in outcomes, τ^2, and the within-cluster variance, σ^2, present in the data). The ICC value is a measure of the correlation as it exists among the correlated observations, see Chapter 2, as noted by Chen and Chen (2021). The value of ICC ranges from 0 (no correlation among clusters) to 1 (perfect correlation among clus-

ters). Mathematically, $\rho_{CC} = \dfrac{\tau^2}{\tau^2 + \sigma^2}$ in a two-level hierarchical design. The larger values of ρ_{ICC} indicate that more variation in outcomes is associated with clustering; that is, a relatively strong relationship exists among clustered individuals in comparison to individuals between clusters. This means individuals (such as patients) within the same cluster (such as a doctor) are more alike on a measured variable than they are alike with individuals in other clusters (i.e., another doctor). This suggests a within-cluster effect is present. Thus, the larger the ICC values, the greater the impact of clustering on a multilevel model. As such, a large ICC value indicates a need to use a multilevel modeling in the data analysis.

In estimating ICC, one estimates both the between-cluster and the within-cluster variances from the data and from their sum. In statistical software, such as SAS/R, estimates of these variances are provided. As such, one can compute the ICC based on those variance estimates. Such variance estimates are demonstrated in the data analysis section later in the chapter.

10.3.2 The Two-Level LME Model

A random-intercept linear mixed-effects model, and a random-intercept and random-slope LME model using the HDP data are now considered. In these analyses, the patients (level-1) are nested within doctors (level-2). This is referred to as the two-level LME models. We present both models.

Random-Intercept Model with Two Levels

Level 1: At level 1, characteristics of the subject i there is outcome y_{il} (such as the patient i in level l in the HDP data), and their outcome is a linear model as follows:

$$y_{il} = \beta_{0l} + \beta_1 x_{ij1} + \beta_2 x_{ij2} + \beta_3 x_{ij3} + \beta_4 x_{ij4} + \beta_5 x_{ij5} + \beta_6 x_{ij6} + \varepsilon_{ij}$$

$\varepsilon_{ij} \sim \mathcal{N}\left(0, \sigma^2\right)$, where β_{0i} is the intercept, and β_k is the coefficient for the predictor x_{ijk} (k = 1 to 6) where $x_{ij1}, \ldots, x_{ij6}$ are the predictors for that subject (such as the patient in this HDP data) j of the ith cluster (such as the doctor in this HDP data) where $j = 1, 2, \ldots, n_i$; n_i is the number of patients within cluster (i.e., doctor) i, and $i = 1, \ldots, n$, is the number of clusters.

Level 2: At level 2, the subject-specific intercept β_{0i} ($i = 1, \ldots n$) depends on the unobserved overall intercept, β_0, and the ith cluster-specific random-effects u_{oi}. Thus,

$$\beta_{0i} = \beta_0 + u_{oi}$$

By successive substitution, the resulting expression is

$$y_{ij} = \beta_0 + \beta_1 x_{ij1} + \beta_2 x_{ij2} + \beta_3 x_{ij3} + \beta_4 x_{ij4} + \beta_5 x_{ij5} + \beta_6 x_{ij6} + u_{oi} + \varepsilon_{ij}.$$

Thus, we have a linear mixed-effects model with random-effects, $u_{oi} \sim \mathcal{N}\left(0, \tau^2\right)$ and $\varepsilon_{ij} \sim \mathcal{N}\left(0, \sigma^2\right)$. This mixed-effects model consists of fixed-effects parameters ($\beta_0, \beta_1, \beta_2, \beta_3, \beta_4, \beta_5, \beta_6$), the between-cluster (i.e., doctor) intercepts variance τ^2, and within-cluster variance σ^2. In the random-intercept LME model, the ICC is the quotient $\tau^2 / \left(\tau^2 + \sigma^2\right)$.

Random-Intercepts and Random-Slopes Model

Level 1: At level 1 the characteristics of the subjects/units and their outcomes are observed. So, at level 1, with slope β_{1i}, the linear model is:

$$y_{ij} = \beta_{0i} + \beta_{1i} x_{ij1} + \beta_2 x_{ij2} + \beta_3 x_{ij3} + \beta_4 x_{ij4} + \beta_5 x_{ij5} + \beta_6 x_{ij6} + \varepsilon_{ij}$$

where $\varepsilon_{ij} \sim \mathcal{N}\left(0, \sigma^2\right)$ represents the within-cluster error distribution. β_{0i} is the cluster-level specific intercept, β_{1i} is the cluster-level specific slope parameter involving x_{ij1}, and β_k (k = 2,3,4,5,6) is the coefficient for the predictor x_{ijk}(k = 2,3,4,5,6).

Level 2: At level 2, the cluster-level specific intercept β_{0i} depends on the unobserved overall intercept β_0, and a cluster-specific error u_{oi},

$$(intercept)\beta_{0i} = \beta_0 + u_{oi}$$

while the β_{1i} is the slope parameter associated with the subject for that i^{th} cluster (such as the doctor in the HDP data), which depends on an overall slope β_1 for the cluster i and a random term u_{1i} as follows:

$$(slope)\beta_{1i} = \beta_1 + u_{1i}$$

By successive substitution, we have the following model:

$$y_{ij} = \beta_0 + \beta_1 x_{1ij} + \beta_2 x_{ij2} + \beta_3 x_{3ij} + \beta_4 x_{4ij} + \beta_5 x_{5ij} + \beta_6 x_{6ij} + u_{oi} + u_{1i} x_{ij1} + \varepsilon_{ij}$$

Thus, a linear mixed-effects model with correlated random-effects is as follows:

$$\begin{pmatrix} \beta_{0i} \\ \beta_{1i} \end{pmatrix} \sim N\left(\begin{pmatrix} \beta_o \\ \beta_1 \end{pmatrix}, \begin{pmatrix} \tau^2, \tau\tau_1\rho_0 \\ \tau\tau_1\rho_0, \tau_1^2 \end{pmatrix} \right)$$

where ρ_0 is the correlation coefficient between the random-intercept and random-slope, and the error term ε_{ij} is assumed to be i.i.d. distributed with a normal distribution and within doctor variance σ^2, i.e., $\varepsilon_{ij} \sim N(0,\sigma^2)$.

This mixed-effects random-intercept and random-slope model has fixed-effects parameters $(\beta_0,\beta_1,\beta_2,\beta_3,\beta_4,\beta_5,\beta_6)$, the between-cluster (i.e., doctor) variances for intercept τ^2 and slope τ_1^2, as well as the correlation coefficient of ρ_0.

10.3.3 The Three-Level LME Model

A three-level LME model has observational units (level-1, such as the patients in the HDP data) that are nested within secondary units (level-2, such as the doctors in the HDP data) and the secondary units are nested within primary units (level-3, such as the hospital in the HDP data). This random intercept and random slope model is described by:

Level-1: At the first level, the observation

$$y_{ijk} = \beta_{0ij} + \beta_{1ij} x_{1ijk} + \beta_2 x_{2ijk} + \beta_3 x_{3ijk} + \beta_4 x_{4ijk} + \beta_5 x_{5ijk} + \beta_6 x_{6ijk} + \varepsilon_{ijk} \qquad (10.1)$$

where i ($i = 1,...,n_i$) denotes the ith primary unit, with j^{th} ($j = 1,...,n_{ij}$) denoting the secondary units and k^{th} ($k = 1,...n_{ijk}$) denotes observational units. The outcome y_{ijk} has predictors x_{pjkl}, where p =1,..., and 6 denotes the number of predictors. The error term ε_{ijk} is assumed to be i.i.d. distributed with a normal distribution and within-cluster variance σ^2, so $\varepsilon_{ijk} \sim N(0, \sigma^2)$.

Level-2 : (Secondary units) At level 2, the random-intercept (β_{0ij}) in the LME model and the random-slope (β_{1ij}) LME model are allowed to vary across secondary units. Then

$$(intercept)\beta_{0ij} = \beta_{0i} + u_{0ij}$$

and

$$(slope)\beta_{1ij} = \beta_{1i} + u_{1ij}.$$

Assume that

$$\begin{pmatrix} \beta_{0ij} \\ \beta_{1ij} \end{pmatrix} \sim N\left(\begin{pmatrix} \beta_{0i} \\ \beta_{1i} \end{pmatrix}, \begin{pmatrix} \tau_{0.2}^2, \tau_{0.2}\tau_{1.2}\rho_{01.2} \\ \tau_{0.2}\tau_{1.2}\rho_{01.2}, \tau_{1.2}^2 \end{pmatrix} \right)$$

Level-3(Primary units):At level 3, the random-intercept (β_{0i}) in the LME model and the random-slope (β_{1i}) LME model are allowed to vary across primary units. Then

$$(intercept)\beta_{0i} = \beta_0 + u_{0i}$$

$$(slope)\beta_{1i} = \beta_1 + u_{1i}$$

where we assume that $\begin{pmatrix} \beta_{0i} \\ \beta_{1i} \end{pmatrix} \sim N\left(\begin{pmatrix} \beta_o \\ \beta_1 \end{pmatrix}, \begin{pmatrix} \tau_{0.3}^2, \tau_{0.3}\tau_{1.3}\rho_{01.3} \\ \tau_{0.3}\tau_{1.3}\rho_{01.3}, \tau_{1.3}^2 \end{pmatrix} \right)$.

In the three-level linear mixed-effects model, there are three additional parameters, including the between-secondary unit variances for intercept $\tau_{0.2}^2$, slope $\tau_{1.2}^2$, as well as the correlation coefficient of $\rho_{01.2}$ between the random-intercept and random-slope. At level-3, there are three additional parameters, including the between-secondary unit variances for intercept $\tau_{0.3}^2$, slope $\tau_{1.3}^2$, as well as the correlation coefficient of $\rho_{01.3}$ between the random-intercept and random-slope. This expansion of the hierarchy can extend to higher levels in a structure, but the challenge is to estimate the parameters in higher dimensions.

10.3.4 Methods for Parameter Estimation

In the analysis of multilevel models, the presence of a correlation structure can be incorporated in the maximum likelihood estimation (MLE). However, due to the random-effects in the models, one usually relies on the restricted maximum likelihood (REML) method as a more appropriate alternative. This approach is extensively discussed in Pinheiro and Bates (2000).

When one uses maximum likelihood estimation, one obtains the estimates of the parameters by maximizing the likelihood of obtaining the given sample. If one attempts to use the maximum likelihood method with multilevel models, then one would formulate a likelihood function based on the LME model and its distribution assumptions at the various levels of the observed data. The resulting likelihood function is maximized through numerical optimization. In particular, the complexity of LME model results in the MLE with no closed-form solution. An iterative numerical optimization algorithm is used to search for parameter values that maximize the likelihood function. As such, the maximum likelihood estimation is computationally intensive and more so for large samples. The MLE is not guaranteed to obtain the parameter estimates, as the search algorithm can diverge without a solution. In addition, MLE is not guaranteed to obtain estimates of the desired parameters due to the possibility of the optimization being local – for most of the search algorithms we implemented. Although, in this situation, different starting values are recommended for the numerical search algorithm.

An alternative is to use the REML estimation. The REML is more accurate in estimating variance parameters as it allows adjustment in the calculation of the degrees of freedom, unlike MLE. The appropriate degree of freedom is important for obtaining REML estimates in variance components, as it accounts for the number of parameters being estimated. The ability of REML to take into account the loss in degrees of freedom due to estimating fixed-effects is key. In contrast, MLE does not account for this loss, and therefore the variance estimators in MLE are biased and underestimated.

Although MLE estimates are biased, this bias may be small with a larger number of high-level clusters. Thus, the difference between MLE and REML becomes very small and negligible as the number of high-level clusters increases. REML is generally the preferred method for estimation in multilevel models; as such, REML is implemented as the default estimation method to obtain variance estimates in statistical software programs.

10.4 Data Analysis Using SAS

10.4.1 Two-Level Random-Intercept LME Model

We fit a model beginning with no covariates:

$$y_i = \beta_0 + \gamma_d + \varepsilon_i$$

Observation   =   Fixed in intercept   Random in doctors   Random in error where β_0 represents the intercept, γ_d represents the random effect in doctors, and ε_i represents the error term. Both γ_d and ε_i are assumed to be normally distributed, $\gamma_d \sim N(0, V_d)$ and $\varepsilon_i \sim N(0, V_e)$

The SAS program for the linear mixed-effects model with random-intercept is:

```
proc mixed data=D2013 covtest noclprint noitprint
method=ml;
class DID;
model tumorsize = / solution;
random-intercept / subject=DID;
run;
```

Model Information	
Dependent Variable	**Tumorsize**
Covariance structure	Variance components
Subject effect	DID
Estimation method	ML
Residual variance method	Profile
Fixed effects SE method	Model-based
Degrees of freedom method	Containment

Dimensions	
Covariance parameters	2
Columns in X	1
Columns in Z per subject	1
Subjects	407
Max obs per subject	40

There are two covariance parameters, one for doctors V_d and the other for random error V_e. There is one fixed-effect term (columns in X) and one random effect term (columns in Z).

Covariance Parameter Estimates					
Cov Parm	**Subject**	**Estimate**	**Standard Error**	**Z Value**	**Pr > Z**
Intercept	DID	$V_d = 29.183$	2.537	11.50	<.0001
Residual		$V_e = 116.26$	1.823	63.76	<.0001

The covariance estimates are given above. The ICC is 29.183/(29.183+116.26) = 20%. 20% of the variance in tumor sizes is attributable to doctors. The random effects for doctors is statistically significant (Z = 11.50, p <.0001).

Solution for Fixed Effects					
Effect	Estimate	Standard Error	DF	t Value	Pr > |t|
Intercept	70.886	0.301	406	235.87	<.0001

The solution gives information about the fixed effect part of the model. A value of 70.889 represents the intercept, as there are no other variables. It has a t-value of 235.87 with p-value <.0001.

The next model fitted is represented by:

$$y_i = \beta_0 + \beta_1 LOS + \gamma_d + \varepsilon_i$$

where β_0 represents the intercept, β_1 is the regression coefficient associated with the variable length of stay, γ_d represents the random effect in doctors, and ε_i represents the error term. Both γ_d and ε_i are assumed to be normally distributed. $\gamma_d \sim N(0,V_d)$ and $\varepsilon_i \sim N(0,V_e)$.

```
Proc mixed data=D2013 covtest noclprint noitprint
method=ml;
class DID; model tumorsize = lengthofstay/ solution;
random-intercept / subject=DID; run;
```

Solution for Fixed Effects					
Effect	Estimate	Standard Error	DF	t Value	Pr > |t|
Intercept	75.795	0.688	406	110.22	<.0001
Length of stay	−0.895	0.113	8117	−7.94	<.0001

The solution gives information about the fixed effect part of the model. A value of −0.894 represents the estimate of the length of stay effect on tumor size. It is statistically significant (t = −7.94, p <.0001). The negative sign associated with length of stay says that length of stay has a negative association with tumor size.

The next model fitted is represented by:

$$y_i = \beta_0 + \beta_1 LOS + \gamma_d + \gamma_{d1} LOS + \varepsilon_i$$

where β_0 represents the intercept, β_1 is the regression coefficient associated with the variable length of stay, γ_d represents the random effect in doctors, γ_{d1} represents the random effects in the slopes associated with LOS, and ε_i represents the error term. The terms γ_d, γ_{d1}, and ε_i are assumed to be normally distributed, $\gamma_d \sim N(0,V_d)$, $\gamma_{d1} \sim N(0,V_{d1})$ and $\varepsilon_i \sim N(0,V_e)$.

```
proc mixed data=D2013 covtest noclprint noitprint
          method=ml;
class DID;
model tumorsize = lengthofstay/ solution;
random-intercept lengthofstay/ subject=DID;
run;
```

Dimensions	
Covariance parameters	3
Columns in X (intercept and LOS)	2
Columns in Z per subject (random intercept and random slope)	2
Subjects	407
Max obs per subject	40

There are three covariance parameters. One for the random intercept, one for the random slope, and one for the within-error variance. There are 2 columns in X, and there are 407 doctors.

Covariance Parameter Estimates					
Cov Parm	Subject	Estimate	Standard Error	Z Value	Pr > Z
Intercept	DID	$V_d = 13.188$	3.994	3.30	0.0005
LengthofStay	DID	$V_{d1} = 0.525$	0.1296	4.05	<.0001
Residual		$V_e = 114.82$	1.805	63.60	<.0001

The covariance estimates are given. The random effects due to differences in doctors is statistically significant (Z = 3.30, p <.0005). Length of stay in terms of tumor size differs statistically significantly across doctors. (Z = 4.05, p-value ≤.0001).

Solution for Fixed Effects					
Effect	Estimate	Standard Error	DF	t Value	Pr > |t|
Intercept	75.788	0.659	406	115.09	<.0001
Length of Stay	−0.893	0.119	405	−7.49	<.0001

The solution gives information about the fixed effects of the model. A value of −0.893 represents that average length of stay and tumor size are negatively correlated at a statistically significant level (t = −7.94, p <.0001).

Covariance Parameter Estimates					
Cov Parm	Subject	Estimate	Standard Error	Z Value	Pr > Z
Intercept	DID	12.225	3.088	3.96	<.0001
Length of Stay	DID	0.611	0.106	5.76	<.0001
Residual		61.649	0.972	63.45	<.0001

Solution for Fixed Effects

Effect	Estimate	Standard Error	DF	t Value	Pr > |t|
Intercept	73.996	1.216	406	60.87	<.0001
Age	−0.115	0.020	7703	−5.63	<.0001
FamilyHx- No	−2.582	0.253	7703	−10.23	<.0001
LengthofStay	0.675	0.110	405	6.14	<.0001
SmokingHx (Current)	5.977	1.888	7703	3.17	0.0016
SmokingHx (Former)	2.926	1.917	7703	1.53	0.1268
CancerStageI	−6.209	0.396	7703	−15.69	<.0001
CancerStageII	−4.343	0.341	7703	−12.74	<.0001
CancerStageIII	−2.592	0.347	7703	−7.46	<.0001
Age*SmokingHx – Current	0.271	0.0381	7703	7.10	<.0001
Age*SmokingHx – Former	0.078	0.038	7703	2.07	0.0380

The solution gives information about the fixed effects of the model. The interaction terms (Age*SmokingHx_Current and Age*SmokingHx-Former) are also listed.

20% of variation in tumor size can be attributed to doctors. The Fixed part of the LME model with random intercept and random slope

> indicates that family history, length of stay, smoking history, and stage of cancer, are negatively correlated with tumor size. Tumor size also differs across the levels of age by smoking history. The random part of the model indicates that tumor size differs across doctors.

10.4.2 Three-Level Random-Slope LME Model Adjusting All Other Covariates

In this section, we illustrate data analysis with three levels of LME to model the tumor size as the response variable. In this model, we use hospital as level 3 and the length-of-stay as the random-slope. The next model fitted is represented by:

$$y_i = \beta_0 + \beta_1 LOS + \gamma_d + \gamma_{d1} LOS + \gamma_h + \gamma_{h1} LOS + \varepsilon_i$$

where β_0 represents the intercept, β_1 is the regression coefficient associated with the variable length of stay, γ_d represents the random effect of doctors within hospitals, γ_{d1} represents the random effects in the slopes associated with doctors within hospitals, γ_h represents the random effects in hospital, γ_{h1} represents the random effects in the slopes associated with hospitals, and ε_i represents the error term. The terms γ_{d1}, γ_h, γ_{h1}, and ε_i are assumed to be normally distributed, $\gamma_d \sim N(0, V_d)$, $\gamma_{d1} \sim N(0, V_{d1})$, $\gamma_h \sim N(0, V_h)$, $\gamma_{h1} \sim N(0, V_{h1})$, and $\varepsilon_i \sim N(0, V_e)$.

```
proc mixed data=D2013 noclprint method=ml;
class DID HID familyHx smokingHx Cancerstage;
model tumorsize = age familyHx lengthofstay smokingHx
        Cancerstage SmokingHx*Age/ solution;
random-intercept lengthofstay/sub=HID type=vc;
random-intercept lengthofstay/sub=DID(HID) type=vc;
run;
```

Covariance Parameter Estimates		
Cov Parm	Subject	Estimate
Intercept	HID	0
LengthofStay	HID	0
Intercept	DID(HID)	12.227
LengthofStay	DID(HID)	0.611
Residual		61.649

Effect	Estimate	Standard Error	DF	t Value	Pr > |t|
Intercept	73.996	1.216	34	60.87	<.0001
Age	−0.115	0.020	7703	−5.63	<.0001
Family history no	−2.582	0.253	7703	−10.23	<.0001
Length of stay	0.675	0.110	34	6.14	<.0001
Smoking current	5.977	1.888	7703	3.17	0.0016
Smoking former	2.926	1.917	7703	1.53	0.1268
Cancer Stage I	−6.209	0.396	7703	−15.69	<.0001
Cancer Stage II	−4.343	0.341	7703	−12.74	<.0001
Cancer Stage III	−2.592	0.347	7703	−7.46	<.0001
Age*Smoking current	0.271	0.038	7703	7.10	<.0001
Age*Smoking- former	0.078	0.038	7703	2.07	0.038

The information about the fixed effects of the model suggests a significant interaction between age and smoking history. The stage of cancer, length of stay, and family history are significant predictors.

The papers by Ene, Leighton, Blue, and Bell (2014) and Ene, Smiley, and Shonenberger (2013) are helpful as additional primers for using SAS (PROC GLIMMIX and PROC MIXED) in multilevel modeling.

10.5 Data Analysis Using R

To model this HDP data using R, we first load the R package "nlme" (i.e., nonlinear linear mixed-effects) as follows:

```
# Call the R library
Library(nlme)
```

10.5.1 Two-Level LME Model with Random-Intercept

We investigate the relationship between "tumor size" and the "Length of Stay" without other covariates. We start with fitting the LME model with random-intercept, which can be implemented in R as follows:

```
# Call "lme" to fit the random-intercept model:
m.int = lme(tumorsize~LengthofStay, random=~1|DID,
        data=hdp)
```

```
# Print the Summary
summary(m.int)

Linear mixed-effects model fit by REML
 Data: hdp
        AIC        BIC      logLik
  65383.81 65412.02 -32687.91
Random-effects:
 Formula: ~1 | DID
         (Intercept) Residual
StdDev:   5.413753 10.74149
```

There are two covariance parameters, one for doctors and the other for random error. The variance of the random effects for doctors is 5.414^2 and the within-variance is 10.741^2.

```
Fixed effects: tumorsize ~ LengthofStay

                Value      Std.Error DF   t-value   p-value

(Intercept)    75.79451 0.6878485 8117 110.19070        0
LengthofStay   -0.89476 0.1127567 8117 -7.93531         0
```

The solution gives information about the fixed effect of the model. Overall, the length of stay is statistically significant in explaining variation in tumor size t = –7.93531 and p-value = 0.

```
Number of Observations: 8525
Number of Groups: 407
```

The number of observations is 8525 (i.e., patients), and the number of groups (level 2) is 407 doctors. The estimated between-doctor standard deviation for random-intercept is $\hat{\tau}$ = 5.414 and within-doctor standard deviation $\hat{\sigma}$ = 10.741. Therefore, the estimated ICC = $\dfrac{5.414^2}{5.414^2 + 10.741^2}$ = 20.26%, which is very high intra-cluster correlation. This suggests that doctor differences explain 20.26% of the variation in tumor sizes. The estimated slope parameter is –0.89476 for the length of stay, which is also statistically significant. This suggests that, regardless of the doctor seen, for every extra day stayed in the hospital, the tumor size from the model is expected to be reduced by 0.89476.

10.5.2 Two-Level LME Model with Random-Intercept and Random-Slope

A LME model with random-slope in length of stay allows the length of stay on tumor size to vary across doctors. This is fitted as follows:

```
# Call "lme" to fit a random-slope model
> m.slope = lme(tumorsize~LengthofStay,
                random=~LengthofStay|DID, data=hdp)
# Print the summary of model fit
> summary(m.slope)

Linear mixed-effects model fit by REML
 Data: hdp
        AIC    BIC    logLik
  65367.32 65409.62 -32677.66
Random-effects:
 Formula: ~LengthofStay | DID
 Structure: General positive-definite, Log-Cholesky
            parametrization
                StdDev      Corr
(Intercept)     3.0662159   (Intr)
LengthofStay    0.6367921   0.352
Residual       10.7213250
```

The solution for the random effects gives a variance estimate of 3.066^2 for the intercept and a variance estimate of 0.637^2 for slopes. A measure of the variance at the observational level (or the estimated within-doctor variance) is $\sigma^{\wedge 2} = 10.721^2$. The correlation coefficient between the random-intercept and random-slope (LengthofStay) $\hat{\rho} = 0.352$.

```
Fixed effects: tumorsize ~ LengthofStay
                Value     Std.Error  DF    t-value     p-value
(Intercept)    75.79284  0.6493677  8117  116.71790      0
LengthofStay   -0.89388  0.1175824  8117   -7.60216      0
```

Number of Observations: 8525
Number of Groups: 407

> The fixed effect of the model suggests that, on average, the length of stay varies statistically significantly across tumor size (t= −7.60216 and p-value = 0). The estimated overall intercept (i.e., overall average tumor size) is 75.793, which is statistically significant with a p-value = 0. The estimated slope parameter is −0.894 for the length of stay, which is also statistically significant. This means that for every extra day stayed in the hospital, tumor size would be predicted from the model to be reduced by 0.894.

10.5.3 Model Selection

To test whether a random-intercept and random-slope LME model is better than the random-intercept LME model, one can make use of the function *"anova"* to use the likelihood ratio test, i.e., chi-square test, as follows:

```
> anova(m.int, m.slope)
     Model     df AIC BIC   logLik     Test        L.Ratio        p-value
m.int    1  4 65383.81  65412.02  -32687.91
m.slop   2  6 65367.32  65409.62  -32677.66   1vs 2 20.49575   <.0001
```

The LME model with random-intercept and random-slope has smaller values of Akaike information criterion (AIC), Bayesian information criterion (BIC), and logLik, as compared to the LME model with random-intercept. It suggests that LME model with random-intercept and random-slope fits the data better. This is consistent with the chi-square likelihood ratio test yielding a likelihood ratio statistic of 20.496, p-value <0.0001.

10.5.4 Three-Level Random-Slope LME Model Adjusting for All Other Covariates

One can fit the three-level LME as described in Section 10.3.3 by simply making the change of the specification of random-effects as *random = ~LengthofStay|HID/DID*, where doctors (i.e., DID) are nested with hospitals (i.e., HID). The R implementation is as follows:

```
# Fit three-level LME Model
mod.tumor3 = lme(tumorsize ~ Age + FamilyHx +
                 LengthofStay
```

```
     + SmokingHx + CancerStage + SmokingHx:Age,
     random=~LengthofStay|HID/DID, data = hdp)
# Print the summary of model fit
> summary(mod.tumor3)

Linear mixed-effects model fit by REML
 Data: hdp
     AIC      BIC      logLik
 60358.6 60485.49 -30161.3
Random-effects:
 Formula: ~LengthofStay | HID
 Structure: General positive-definite, Log-Cholesky
                                      parametrization
              StdDev         Corr
(Intercept)  0.0132620072   (Intr)
LengthofStay 0.0005410812   -0.002
 Formula: ~LengthofStay | DID %in% HID
 Structure: General positive-definite, Log-Cholesky
                                      parametrization
              StdDev       Corr
(Intercept)  2.6161383    (Intr)
LengthofStay 0.6586244    0.584
Residual     7.8664149


Fixed effects: tumorsize ~ Age + FamilyHx + LengthofStay
                          + SmokingHx
+ CancerStage + SmokingHx:Age
```

	Value	Std.Error	DF	t-value	p-value
(Intercept)	71.17688	1.6268071	8108	43.75250	0.0000
Age	0.15639	0.0336784	8108	4.64351	0.0000
FamilyHxyes	2.57214	0.2525196	8108	10.18592	0.0000
LengthofStay	0.67286	0.1073684	8108	6.26686	0.0000
SmokingHxformer	-3.04753	2.2910532	8108	-1.33019	0.1835
SmokingHxnever	-5.96136	1.8878107	8108	-3.15782	0.0016
CancerStageII	1.86949	0.2359693	8108	7.92261	0.0000
CancerStageIII	3.62474	0.3037252	8108	11.93427	0.0000
CancerStageIV	6.20802	0.3959717	8108	15.67793	0.0000
Age:SmokingHxformer	-0.19266	0.0463150	8108	-4.15971	0.0000
Age:SmokingHxnever	-0.27119	0.0381189	8108	-7.11436	0.0000

> The solution gives information about the fixed effects of the model. The interaction terms (Age*SmokingHx_Current and Age*SmokingHx-Former) are also listed.

> 20% of variation in tumor size can be attributed to doctors. The fixed part of the LME model with random intercept and random slope indicates that family history, length of stay, smoking history, and stage of cancer are negatively correlated with tumor size. Tumor size also differs across the levels of age by smoking history, as indicated by the significant interactions.

Number of Observations: 8525

Number of Groups:

HID DID %in% HID

35   407

From the output, the "Number of Observations" is 8525, which are nested within 407 doctors (i.e., DID %in% HID) and 35 hospitals (i.e., in "HID"). The estimated random-effects at level-2 is displayed as follows:

```
Formula: ~LengthofStay | DID %in% HID
Structure: General positive-definite, Log-Cholesky
                                     parametrization
                 StdDev        Corr
(Intercept)    2.6161383      (Intr)
LengthofStay   0.6586244      0.584
Residual       7.8664149
```

That is, in this HDP, the estimated level-2 standard deviations are $\tau_{0.2} = 2.616$ for random-intercept, and $\tau_{1.2} = 0.659$ for random-slope with correlation coefficient $\rho_{01.2} = 0.584$. Similarly, at level-3 (i.e., hospital), the output is displayed as follows:

```
Formula: ~LengthofStay | HID
 Structure: General positive-definite, Log-Cholesky
                                     parametrization
                 StdDev           Corr
(Intercept)    0.0132620072      (Intr)
LengthofStay   0.0005410812      -0.002
```

That is, in this HDP, the estimated level-3 standard deviations are $\tau_{0.3} = 0.0132620072$ for random-intercept and $\tau_{1.3} = 0.0005410812$ for random-slope with correlation coefficient $\rho_{01.3} = -0.002$. These estimates are very small since the "tumorsize" was simulated to vary only for level-2. The significance of level-3 LME model to level-2 LME model can be simply tested as follows:

```
> anova(mod.tumor, mod.tumor3)
                Model df      AIC       BIC        logLik  Test    L.Ratio
p-value
mod.tumor       1 15 60352.6  60458.34 -30161.3
mod.tumor3      2 18 60358.6  60485.49 -30161.3  1 vs 2  3.416037e-05 1
```

This indicates again that the level-2 LME model is virtually identical to the level-3 LME model with the chi-square test of p-value of 1.

10.5.5 Three-Level LME Model with Random-Intercept and Random-Slope

In this section, an extension to three-level LME models is presented. In this demonstration, hospital is used as level-3 random-intercept and the length of stay as slope, also at level-3. We fit a LME model with random-intercepts and random-slopes to three-level hierarchical data:

```
# Call "lme" to fit LME model with other covariates
>mod.tumor3 = lme(tumorsize ~ Age + FamilyHx +
LengthofStay
         + SmokingHx + CancerStage + SmokingHx:Age,
         random=~LengthofStay|HID/DID, data = hdp)
# Print the model fitting
>summary(mod.tumor)

Linear mixed-effects model fit by REML
   Data: hdp
            AIC          BIC          logLik
            60358.6      60485.49     -30161.3

Random-effects:
 Formula: ~LengthofStay | HID
 Structure: General positive-definite, Log-Cholesky
                                      parametrization
              StdDev          Corr
(Intercept    0.0132620072    (Intr)
LengthofStay  0.0005410812    -0.002

 Formula: ~LengthofStay | DID %in% HID
 Structure: General positive-definite, Log-Cholesky
                                      parametrization
              StdDev          Corr
(Intercept)   2.6161383       (Intr)
LengthofStay  0.6586244       0.584
Residual      7.8664149
```

There are five covariance parameters. An intercept and slope at levels 2 and 3 and the variance among the observations at level 1. There are 15 parameters in the fixed part of the model (the main effects and the interaction plus the intercept).

```
Fixed effects: tumorsize ~ Age + FamilyHx + LengthofStay + SmokingHx +
CancerStage +  SmokingHx:Age
                     Value Std.Error  DF t-value p-value
(Intercept)       71.17688 1.6268071 8108 43.75250  0.0000
Age                0.15639 0.0336784 8108  4.64351  0.0000
FamilyHxyes        2.57214 0.2525196 8108 10.18592  0.0000
LengthofStay       0.67286 0.1073684 8108  6.26686  0.0000
SmokingHxformer   -3.04753 2.2910532 8108 -1.33019  0.1835
SmokingHxnever    -5.96136 1.8878107 8108 -3.15782  0.0016
CancerStageII      1.86949 0.2359693 8108  7.92261  0.0000
CancerStageIII     3.62474 0.3037252 8108 11.93427  0.0000
CancerStageIV      6.20802 0.3959717 8108 15.67793  0.0000
Age:SmokingHxformer -0.19266 0.0463150 8108 -4.15971  0.0000
Age:SmokingHxnever  -0.27119 0.0381189 8108 -7.11436  0.0000
```

The information about the fixed effects of the model suggests a significant interaction between age and smoking history. The stage of cancer, length of stay, and family history are significant predictors. From the output of this random-slope LME model, the estimated within-doctor standard deviation is now reduced to $\hat{\sigma} = 7.866$. The estimated between-doctor standard deviation for random-intercept is $\hat{\tau}_0 = 2.618$, for random-slope is $\hat{\tau}_0 = 0.658$, and the correlation coefficient is $\hat{\rho} = 0.83$. The estimated overall intercept (i.e., overall tumor size) is 71.177, which is statistically significant with a p-value = 0. There is a statistically significant interaction between "Age" and "SmokingHx". All "CancerStage" and "FamilyHx" are statistically significant which further confirms the conclusions from Chapters 4 and 5.

10.6 Discussions and Comments

For model diagnostics, a plot of the residual QQ-plot as shown in Figure 10.3 is useful. The QQ-plot checks to see if the residuals from this LME model support a normal distribution, which is very supportive in this data analysis.

The models in this chapter are subject-specific. They tell about the mean given certain conditions as opposed to the marginal models as we did in Chapter 9. It is not a new phenomenon that a statistical model should reflect

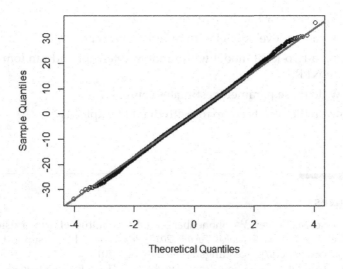

FIGURE 10.3
Residual QQ-Plot

the design and the method whereby the data are collected. Fitting models that reflect the different levels of the data as collected are important in analyzing such data. Such models are often referred to as clustered or longitudinal data models. Longitudinal data are obtained from the same subject/unit at different points in time. There is variation due to the correlation of observations from the same subject over time.

Clustered or hierarchical data consist of the observations structured at different levels. Thus, there are multiple sources of variation in hierarchical data. The standard linear regression model ignores several sources of variation from different units. While the hierarchical linear regression model incorporates these different sources of variation.

The statistical programs outlined in this chapter provide a practical guide for evaluating nested three-level models. These models allow the combination of covariates at different levels, within level and across level interaction terms and random components.

10.7 Exercises

Use the Medicare data to fit the following:

1. Use the length of stay as the response with covariates as NDX, NPR, DX101, and DX203, with patient as the second-level variable. Fit a subject-specific model. Time is nested in patient as a factor to:

    a. Fit a standard logistic regression model.

    b. Fit a two-level model with random intercept.

    c. Fit a two-level model with random intercept and random slope in NPR.

2. How do these parameter estimates compare?

3. How do the results compare with that in Chapter 11?

References

Bell, B.A., Ene, M., Smiley, W., Shonenberger, J.A.: A multilevel primer using SAS®. *PROC MIXED. SAS Global Forum 2013 Proceedings.* http://support.sas.com/resources/papers/proceedings13/433-2013.pdf (2013).

Chen, D.G., Chen, J.K.: *Statistical Regression Analysis Using R: Longitudinal and Multi-Level Modeling.* Springer, New York (2021)

Ene, M., Leighton, E.A., Blue, G.L., Bell, B.A.: Paper 3430-2015 multilevel models for categorical data using SAS®. PROC GLIMMIX: The Basics. https://support.sas.com/resources/papers/proceedings15/3430-2015.pdf (2014)

Pinheiro, J.C., Bates, D.M: *Mixed-Effect Models in S and Splus.* Springer, New York (2000)

11

Modeling Correlated Binary Outcomes Through Hierarchical Logistic Regression Models

11.1 Research Interest/Question

In the standard logistic regression model, one assumes that the observations in a model fit are independent. However, when one encounters observations such as in a longitudinal study or when clustering is present, one has to provide a method that addresses the inherent correlation. More importantly, if one uses the standard logistic regression model on such correlated observations, it leads to underestimates of the standard errors of the regression coefficients and may lead one to conclude significant regression coefficients when, in fact, there is not a significant relationship.

In longitudinal data, the observations are obtained from the subjects or units followed over time and/or repeatedly under different experimental conditions, and these observations are correlated. In clustered data, the subjects are observed in groups or families or census tracts or neighborhoods, or clusters. Often times, such data originate from a hierarchical structure consisting of units and subunits at different levels. Hierarchical structures create intra-class correlation (ICC) at different levels. Such intra-class correlation is often modeled through random effects.

In standard logistic regression models, the focus is on modelling marginal means and the interpretation pertains to the population average. However, in this chapter as the presence of correlation is evident, the emphasis is on modeling the conditional mean. The conditioning is done on the random effects. It sounds nonintuitive as random effects are the unobservable differential effects among clusters. In particular, this chapter provides interpretation that relates to subgroups as opposed to overall averages. It focuses on subject-specific models for binary outcomes.

11.2 Bangladesh Data

Bangladesh has one of the densest populations in the world. In the early 1950s, family planning was introduced by the government of Bangladesh as a means of controlling fast-rising population growth. As a means of birth control, contraceptive use is an effective method to help in reducing the fertility rate. It is highly encouraged by the government of Bangladesh, with the expectation of a decline in the fertility rate in the long term (Khan and Shaw 2011). It is widely believed that the decline in fertility rate is associated with the prevalence of contraceptive use. In addition, it is reasonable to want to understand how the efforts in family planning had differential effects in some areas. In the 2011 Bangladesh Demographic and Health Survey (BDHS), data regarding contraceptive use of 16,186 women is analyzed. The survey design consists of a three-level hierarchical data structure. These women are nested in the specific geographic areas (600 areas), and the areas are nested in the major divisions (seven divisions) in Bangladesh. The women are either "never married or currently nonpregnant women" aged 13–49 as shown in Figure 11.1.

In this chapter, the binary response variable of interest is contraceptive use (CUC) coded as "1" and zero denoting a woman that did not use contraceptives. The covariates consist of current age, age at first marriage, and the number of living children. In addition, there are binary predictors, including whether the household had a radio (*Radio*), whether the household had a television (*Television*), plus categorical predictors, such as the highest educational level of the female respondent (*Education*), place of residence of the female respondent (*Home*), religious belief (religion), and family wealth (*Wealth Index*). A subset of these data is given in Table 11.1.

The primary interest of the analysis is to determine the impact of certain characteristics, geographic areas, and divisions on the likelihood of contraceptive use. Therefore, a woman is at the observational level (level-1), geographic area is at the secondary level (level-2), and division at the highest level (level-3). This three-level hierarchical structure is used to investigate the relationship between contraceptive use and their characteristics as well as taking random effects into consideration. The specific research questions are:

Q1. How much variance in contraceptive use is attributable to geographical districts and divisions?

Q2. Does the influence of age vary among geographical areas?

Q3. What is the impact of geographic area, while controlling for the characteristics of the woman?

Q4. Do the divisions have an impact on contraceptive use while controlling for the characteristics of the woman?

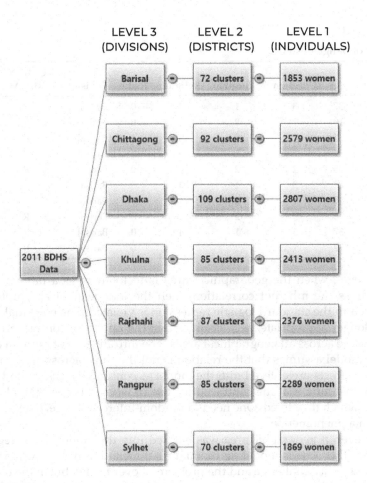

FIGURE 11.1
Hierarchical structure of the 2011 BDHS data

11.3 Hierarchical Models

Define the probability that the ith woman in the jth area of the kth division uses contraceptive as $P(CUC = 1) = \pi_{ijk}$. Consider the probability π_{ijk} with covariates age, children, education, and religion such that the systematic component is

$$g(\pi_{ijk}) = \beta_0 + \beta_1 age + \beta_2 children + \beta_3 education + \beta_4 religion,$$

where β_p for $p = 1,2,...4$ are regression coefficients, and g is a link function (Dobson 2002). If one uses this systematic function, then one has a marginal model, as in Chapter 7. This marginal model provides an opportunity to model the marginal probability with regards to age, children, education, and religion.

Statistical Analytics for Health Data Science with SAS and R

TABLE 11.1

Subset of Bangladesh Data

District	Age	Children	Education	Religion	Radio	Divisions	URBAN/ RURAL	CUC
1	37	3	1	1	0	Barisal	R	0
1	19	1	2	1	0	Barisal	R	0
1	25	2	2	1	1	Barisal	R	1
1	30	2	1	1	0	Barisal	R	1
1	25	2	2	1	0	Barisal	R	1
1	31	3	1	1	0	Barisal	R	1
1	23	1	2	1	0	Barisal	R	1
1	36	3	2	1	0	Barisal	R	1
1	37	3	1	1	0	Barisal	R	1
1	22	1	2	1	0	Barisal	R	1

However, when the geographical area is factored in as a random effect to address the inherent correlation, then the independent assumption as required in the standard logistic regression is violated. The marginal model is no longer appropriate, as it assumes that the impact of contraceptive use is the same across all geographical areas. The introduction of random effects in the model assumes that the relation is not the same across all areas. The random effects are entered into the model as continuous. It is a conditional model, meaning that the contraceptive use of women is conditional on the area in which they lived. One needs a random-intercept model to extrapolate in such a phenomenon.

However, if geographic area was entered into the model as a categorical variable, then one is modeling the marginal mean but has to overcome the interpretability challenge and the problem of over fitting, but at the expense of the correlation – which is not factored in. Moreover, the results are only applicable to the categories in the data.

11.3.1 Two-Level Nested Logistic Regression with Random-Intercept Model

A random-intercept model is the simplest of the correlated models or generalized linear mixed models. This model has intercept the geographical area, as in our example. In other words, the random-effects model accounts for any unaccounted variation beyond the covariates in the data. The random-intercept model is:

$$\text{logit}\{\mu_{ij} \mid X_{ij1}...X_{ijP}, \gamma_j\} = \beta_0 + \beta_1 X_{ij1} + ...+ \beta_P X_{ijP} + \gamma_j$$

$$\gamma_j \sim \text{Normal}(0, \delta_\gamma^2)$$

where γ_j has mean zero, β_m (m = 1,...p) represents the regression coefficients, and the parameter δ_γ^2 indicates the variance in the population of random effects γ_j. Thus, it measures the degree of heterogeneity among geographical areas (Blackwelder 1998). As such, the model addresses the conditional distribution of the probability of outcome given the random effects (due to area). Thus, it models the conditional mean of a binary response given the geographical area (random effects).

The random-effects model can be thought of as augmenting the linear predictor in a generalized linear model [$\text{logit}\{\mu_{ij} \mid X_{ij1}...X_{ijP}, \gamma_j\} = \beta_0 + \beta_1 X_{ij1} +.....$ $+\beta_P X_{ijP}$] with the random effect γ_j for unit j (at the level-2, i.e., j^{th} area). It represents the unmeasured covariates. In particular, it models heterogeneity in the data. The random effect (level-2 intercept) addresses heterogeneity due to the geographical area j as j goes from 1 to J, representing the set of area. It allows each of the 600 areas to have a different intercept but assumes the same slope. In the Bangladesh data, these random effects represent the influence of the differences between areas on contraceptive usage that are not captured by the covariates (fixed effect) in the data. In other words, the random-effects model accounts for any unaccounted variation beyond the covariates in the data.

The random intercept model or generalized linear mixed model (GLMMIX) can be written as:

$$g\left[E\left(y_{ij} \mid \gamma_j\right)\right] = g\left(\mu_{ij}\right) = \beta_0 + \overbrace{\beta_1 X_{ij1} +.....+ \beta_P X_{ijP}} + \overbrace{\gamma_j} ,$$

where the function g is the link function that is twice differentiable. In this case, *the g* function is the logit. The random-intercept model makes use of two distributions, one distribution on the observations and a second distribution at the next level on the random effects. In this example, there is a Bernoulli distribution on observations, (i.e. the conditional response $y_{ij} \mid \gamma_j$) and a normal distribution for the random effects γ_j.

In these generalized linear mixed models, we refer to an R-side and a G-side. The R-side denotes all of the random components, while the G-side denotes the fixed effects (Goldstein, 2003). The G-side effects make up the link component and are thus interpretable as:

$$\{\beta_0 + \beta_1 X_{ij1} +.....+ \beta_P X_{ijP}\}.$$

The R-side represents distribution of the random effects and the distribution of the unaccounted variation. The generalized linear mixed model is referred to as a subject-specific model, it models $E\left(y_{ij} \mid \gamma_j\right)$. It differs from modeling the unconditional expectation $E\left(y_{ij}\right)$ (Dean and Nielsen 2007; Have, Kunselman and Tran 1999).

11.3.2 Interpretation of Parameter Estimates in a Subject-Specific Model

An important consideration for the generalized linear mixed model is the interpretation of the unobserved γ_j.

$$g\left[E\left(y_{ij} \mid \gamma_j\right)\right] = g\left(\mu_{ij}\right) = \beta_0 + \overbrace{\beta_1 X_{ij1} + \ldots + \beta_P X_{ijP}} + \widetilde{\gamma_j}$$

- Interpret β_0 as the log-odds of $y_{ij} = 1$ when $X_{ijk} = 0$ (k=1, ..., P) and $\gamma_i = 0$;
- When $\gamma_j = 0$ for all j, then there is no need to model the random effects as there is no differential effect.
- β_k is the effect on the log-odds for a unit increase in X_{ijk} for individuals in the same group, which will have the same random effect or grouping value of γ_j. The term γ_j is the effect of being in group j.
- The $[\beta_1 \ldots, \beta_P]$ parameters are measures of the overall effect without taking into consideration levels of the random effect.

Larsen et al. (2000) discussed that conditioning on the random effects yields similar interpretations in terms of odds ratios, as in the case for standard logistic regression models. However, it is not necessarily possible to condition on unobservable random effects. Then, the odds ratio is a random variable rather than a fixed parameter. This important aspect should be kept in mind when interpreting the model. Observations within a cluster are assumed independent given the random cluster effect (conditional independence).

11.3.3 Statistical Analysis of DATA Using SAS Program

A two-level logistic regression model with random-intercepts is fitted to the Bangladesh data with the SAS program as in Schabenberger (2005). The response is contraceptive use. The fixed effects in the model are: age of the woman, number of living children, level of education, and whether or not the woman has religious beliefs. The areas, seen as clusters, are random effects. The model

$$\text{logit}\{\mu_{ij} \mid X_{ij1} \ldots X_{ijP}, \gamma_j\} = \beta_0 + \beta_1 X_{ij1} + \ldots + \beta_P X_{ijP} + \gamma_j$$

treats X_{ijk} (k=1, ...P) as fixed and γ_j as random.

```
PROC GLIMMIX DATA=BANGLADESH ;
CLASS CLUSTER;
MODEL CUC(EVENT='1') = AGE CHILDREN EDUCATION RELIGION/DIST=BINARY
                    LINK=LOGIT DDFM=BW SOLUTION OR;
RANDOM INTERCEPT/SUBJECT =CLUSTER SOLUTION TYPE=VC;
COVTEST/WALD; RUN;
```

These SAS statements will fit this generalized linear mixed model with logit link.

The GLIMMIX Procedure	
Model information	
Dataset	WORK.BANGLADESH
Response variable	CUC
Response distribution	Binary
Link function	Logit
Variance function	Default
Variance matrix blocked by	cluster
Estimation technique	Residual PL
Degrees of freedom method	Between-Within

The response variable is CUC. This is binary and has a Bernoulli distribution for each woman. The logit is the log of {(probability of CUC = 1)/ (probability of CUC = 0)}.

Class Level Information		
Class	Levels	Values
Cluster	600	1 2 3 4 5 6 7 8 9 10 11 12 13 14 15 16 17 18 19 20 21 22 23 24 25 26 27 28 29 .... 598 599 600
Number of observations read	16186	
Number of observations used	16186	

There are 600 clusters. The total number of women in the data is 16,186.

Response Profile		
Ordered Value	CUC	Total Frequency
1	0	6508
2	1	9678

The GLIMMIX procedure is modeling the probability that CUC = '1'.

It is important to know what is modeled. It is probability of CUC = 1 as opposed to CUC = 0. This gives the order of the ratio in the logit and hence the direction with interpretation of the covariates.

$$\text{logit}\{\mu_{ij} \mid X_{ij1}...X_{ijP}, \gamma_j\} = \beta_0 + \beta_1 X_{ij1} + ... + \beta_P X_{ijP} + \gamma_j$$

Dimensions	
G-side Cov. parameters	1
Columns in X	5
Columns in Z per subject	1
Subjects (blocks in V)	600
Max obs per subject	39

The generalized linear mixed model has two parts. The fixed effects, G-side, and the random effects, R-side. The random effects is the cluster (geographical area). The variance is of random effects in the G-side covariate parameter. The columns in X are intercepts: age, children, education, and religion. The maximum cluster size is 39.

			Iteration History		
Iteration	Restarts	Subiterations	Objective Function	Change	Max Gradient
0	0	3	71501.229333	0.14148554	0.001178
1	0	2	70727.151893	0.00945525	0.000011
2	0	1	70731.871987	0.00006452	3.838E-6
3	0	1	70731.943142	0.00000138	1.759E-9
4	0	1	70731.944588	0.00000003	1.65E-8
5	0	0	70731.944617	0.00000000	4.492E-7

Convergence criterion (PCONV=1.11022E-8) is satisfied.

The convergence is satisfied. This is useful to know, as the coefficients are obtained through an iterative process. This may not always converge. Any result, given if the convergence criterion is not met, cannot be trusted.

Fit Statistics	
−2 Res Log pseudo-likelihood	70731.94
Generalized chi-square	15770.86
Gener. chi-square/DF	0.97

The fit statistic gives an idea of how well the model fits. The ratio Gener. Chi-Square DF should be close to one as a good fit.

Covariance Parameter Estimates					
COV PARM	SUBJECT	ESTIMATE	STANDARD ERROR	Z VALUE	PR > Z
Intercept	cluster	0.2931	0.02790	10.51	<.0001

The variance of random effects is 0.2931. It is significant (p <.0001). Therefore, accounting for the random effects (geographical area) in the model is a good decision.

Solutions for Fixed Effects					
Effect	Estimate	Standard Error	DF	t Value	Pr > |t|
Intercept	0.6296	0.09974	599	6.31	<.0001
age	−0.05428	0.002457	15582	−22.09	<.0001
children	0.3894	0.01512	15582	25.76	<.0001
education	0.1670	0.02093	15582	7.98	<.0001
religion	0.3217	0.05771	15582	5.57	<.0001

The four covariates are significant (p <.0001) in the model. They are fixed effects. They tell about the relations in the data. Age has a negative sign, so older women were less likely to use contraceptives. However, those with more children, higher level of education, and having religious beliefs are more likely to use contraceptives. We say this since the signs on the coefficients are positive.

Odds Ratio Estimates										95% Confidence Limits	
Age	Children	Education	Religion	Age	Children	Education	Religion	Estimate	DF		
32.42	2.402	1.230	1.123	31.421	2.402	1.2295	1.123	0.947	15582	0.943	0.952
31.42	3.402	1.230	1.123	31.421	2.402	1.2295	1.123	1.476	15582	1.433	1.520
31.42	2.402	2.230	1.123	31.421	2.402	1.2295	1.123	1.182	15582	1.134	1.231
31.42	2.402	1.230	2.123	31.421	2.402	1.2295	1.123	1.380	15582	1.232	1.545

This table provides estimates of the odds ratio depending on certain covariate values. Note the difference represents an increase by one unit.

Solution for Random Effects						
Effect	Subject	Estimate	Std Err Pred	DF	t Value	Pr > |t|
Intercept	cluster 1	0.2445	0.3442	16181	0.71	0.4775
Intercept	cluster 2	0.5747	0.3348	16181	1.72	0.0861
Intercept	cluster 3	0.6731	0.3427	16181	1.96	0.0495
Intercept	cluster 4	0.5380	0.3456	16181	1.56	0.1196
Intercept	cluster 5	0.5876	0.3490	16181	1.68	0.0923
		...	...	...	...	...
Intercept	cluster 599	−0.02973	0.3229	16181	−0.09	0.9266
Intercept	cluster 600	−0.4422	0.3240	16181	−1.36	0.1723

> A measure of the random effects for each cluster is provided. We provide a truncated version of the 600 clusters. We note that cluster 3 does provide a significant though mild effect compared to others.

11.3.4 Statistical Analysis of DATA Using R Program

A two-level logistic regression model with random-intercept is fitted to the Bangladesh data with the R program. The response is contraceptive use. The fixed effects in the model are: age of the woman, number of living children, level of education, and whether or not the woman has religious beliefs. The areas (clusters) are random effects.

```
# load the data into R
Bangladesh<-read.csv(("Data File path/Bangladesh.csv"))
attach(Bangladesh)


Correlated Logistic Regression - Two Levels with
            Random-Intercept
## load the R library
library(lme4)
## Loading required package: Matrix
## Define the control for optimization
newcont = glmerControl(
      check.conv.grad = .makeCC("ignore", tol = 1e-2,
                      relTol = NULL),
      check.conv.singular = .makeCC(action =
                      "ignore", tol = 1e-4),
      check.conv.hess   = .makeCC(action = "ignore", tol
                      = 1e-4))
```

```
## Call "glmer" to fit the random-intercept model
glmer.out=glmer(CUC~age+children+education+religion+(1
                |cluster),
data=Bangladesh,family=binomial,control=newcont)
summary(glmer.out)

## Generalized linear mixed model fit by maximum
##              likelihood (Laplace
##   Approximation) [glmerMod]
## Family: binomial ( logit )
## Formula: CUC ~ age + children + education + religion
##          + (1 | cluster)
##   Data: Bangladesh
## Control: newcont
##
##      AIC      BIC   logLik   deviance  df.resid
##   20749.3  20795.5 -10368.7   20737.3     16180
##
## Scaled residuals:
##     Min       1Q    Median      3Q        Max
##   -4.1669  -1.0002   0.5707   0.7691    3.0007
##
## Random effects:
##    Groups  Name          Variance  Std.Dev.
##    cluster (Intercept)    0.3088    0.5557
## Number of obs: 16186, groups: cluster, 600
##
```

> The variance of random effects is 0.3088 with a standard deviation of
> 0.5557, which is slightly different from 0.2931 in the SAS due to different
> default settings on fitting methods.

```
## Fixed effects:
##                Estimate Std. Error  z value  Pr(>|z|)
## (Intercept)    0.648731   0.101454    6.394   1.61e-10 ***
## age           -0.055780   0.002542  -21.941   < 2e-16  ***
## children       0.400118   0.015820   25.292   < 2e-16  ***
## education      0.171058   0.021204    8.067   7.20e-16 ***
## religion       0.329493   0.058464    5.636   1.74e-08 ***
## ---
## Signif. codes: 0 '***' 0.001 '**' 0.01 '*' 0.05 '.' 0.1 ' ' 1
##
```

The four covariates are significant (p <.0001) in the model. They are fixed effects. They tell about the relations in the data. Age has a negative sign, so older women were less likely to use contraceptives. However, those with more children, higher level of education, and having religious beliefs are more likely to use contraceptives. We say this since the signs on the coefficients are positive.

11.4 Two-Level Nested Logistic Regression Model with Random-Intercept and Random-Slope

In the two-level nested logistic regression model with random-intercepts in Section 11.3, it was assumed that the rate of change remains the same for each area (cluster). Therefore, in the Bangladesh data, one is assuming the 600 areas are represented by parallel planes.

Let us consider a two-level logistic regression model with random-intercept $[\gamma_{0j}]$ and random-slope $[\gamma_{1j}]$:

$$\text{logit}(P_1 \mid \gamma_{0j}, \gamma_{1j}) = \beta_0 + \beta_1 X_1 + \beta_2 X_2 + \gamma_{0j} + \gamma_{1j} Z_1$$

with Z_1 is a measurable variable associated with random-slope where γ_{0j} is distributed as normal with a mean of zero and the variance $\delta_{\gamma 0}^2$ and γ_{1j} is distributed as normal with a mean of zero and variance $\delta_{\gamma 1}^2$ (Schabenberger 2005). This model assumes that, given γ_{0j} and γ_{1j}, the responses from the same cluster are mutually independent or, rather, that the correlation between units from the same cluster is completely explained by them having been in the same cluster. As such, these are called subject-specific parameter models (Hu, Goldberg, Hedeker, Flay, and Pentz 1998). Each cluster has its own intercept and its own slope.

The random-intercept model assumes that those intercepts have a normal distribution with mean zero and variance $\delta_{\gamma 0}^2$. The random-intercept model assumes that each cluster has a different point at baseline but changes at the same rates. However, if $\delta_{\gamma 0}^2$ is found to be different from zero, then one can conclude that there is a need for assuming different intercepts.

Similarly, one assumes that in a random-slope model the rate of change based on a particular cluster has a normal distribution with mean zero and variance $\delta_{\gamma 1}^2$. However, if $\delta_{\gamma 1}^2$ is found to be different from zero, then one can conclude that there is a need for assuming different slopes.

11.4.1 Statistical Analysis of DATA with SAS Program

A two-level logistic regression model with random-intercept and random-slope is fitted to the Bangladesh data using the SAS program. The response is contraceptive use. The fixed effects in the model are: age of the woman, number of living children, level of education, and whether or not the woman has religious beliefs. The areas (clusters) are the random effects and age is the random-slope, assuming that the relationship between the covariate age to CUC are geographical area-specific. Thus the model is

$$\text{logit}(P_1 \mid \gamma_{0j}, \gamma_{1j}) = \beta_0 + \beta_1 X_1 + \beta_2 X_2 + \gamma_{0j} + \gamma_{1j} Z_1$$

These SAS statements will fit this generalized linear mixed model (random intercepts and random slopes with logit link are given).

11.4.2 Statistical Analysis of DATA Using SAS Program

```
**Two levels with random-intercept and slope;
PROC GLIMMIX data=Bangladesh ;
CLASS cluster;
model CUC(event='1')= age children education religion/
DIST=BINARY LINK=LOGIT DDFM=BW SOLUTION OR;
random intercept age/subject =cluster solution TYPE=VC;
covtest/WALD;
run;
```

Selected The GLIMMIX Procedure

Dimensions

G-side Cov. parameters	2
Columns in X	5
Columns in Z per subject	2
Subjects (blocks in V)	600
Max obs per subject	39

The generalized linear mixed model has two parts: the fixed effects and the random effects. In this case, the random effects is the cluster with a certain variance and the slope with a certain variance. The variance is noted by the G-side covariance parameters. The columns in X are intercepts: age, children, education, and religion. The largest number of women in a cluster is 39.

Covariance Parameter Estimates

Cov Parm	Subject	Estimate	Standard Error	Z Value	Pr > Z
Intercept	cluster	0.2324	0.04180	5.56	<.0001
age	cluster	0.000066	0.000037	1.80	0.0360

The variance of random effects (cluster) is 0.2324. It is significant (p <.0001). Therefore, accounting for the random effects in the model is a good decision. The random-slope has variance 0.000066, which is also significant (p = 0.036).

		Solutions for Fixed Effects			
Effect	Estimate	Standard Error	DF	t Value	Pr > |t|
Intercept	0.6215	0.09941	599	6.25	<.0001
age	−0.05449	0.002492	15582	−21.87	<.0001
children	0.3938	0.01522	15582	25.88	<.0001
education	0.1692	0.02099	15582	8.06	<.0001
religion	0.3223	0.05793	15582	5.56	<.0001

The four covariates are significant (p <.0001) in the model. They are fixed effects. They tell about the relations in the data. Age has a negative sign, so older women were less likely to use contraceptives. However, those with more children, higher level of education, and having religious beliefs are more likely to use contraceptives. We say this since the signs on the coefficients are positive.

11.4.3 Statistical Analysis of Data Using R Program

A two-level logistic regression model with random-intercept and random-slope is fitted to the Bangladesh data with the R program. The response is contraceptive use. The fixed effects in the model are: age of the woman, number of living children, level of education, and whether or not the woman has religious beliefs. The geographical areas (clusters) are the random effects, and age is the random-slope.

```
## Read in the data
Bangladesh<-read.csv(("Data file path/Bangladesh.csv"))
attach(Bangladesh)
```

Correlated Logistic Regression: Two Levels with Random-Intercept and Random-Slope

```
## Call "glmer" to fit the random-intercept and random-
                slope model
glmer.out2=glmer(CUC~age+children+education+religion+(
                age|cluster),
```

```
data=Bangladesh, family=binomial, control=newcont)
summary(glmer.out2)

## Generalized linear mixed model fit by maximum
                 likelihood (Laplace
##        Approximation)    [glmerMod]
##      Family: binomial   ( logit )
## Formula: CUC ~ age + children + education + religion
               + (age | cluster)
##      Data: Bangladesh
## Random effects:
## Groups     Name          Variance      Std.Dev.     Corr
## cluster    (Intercept)   0.8746545     0.93523
##            age           0.0006065     0.02463      -0.80
## Number of obs: 16186, groups:     cluster,      600
##
```

The variance of random effects (cluster) is 0.8746 with standard deviation 0.93523. The random-slope has variance 0.00060605 with standard deviation 0.02463. Notice the difference of these estimates to those from SAS and STATA, which is due to different methods of parameter estimation. In R, the restricted maximum likelihood (REML) is used as the default, but in SAS and Stata, the maximum likelihood estimation (MLE) is used as default.

```
## Fixed effects:
##                 Estimate Std. Error     z value Pr(>|z|)
## (Intercept)     0.656140    0.107545     6.101   1.05e-09 ***
## age            -0.056692    0.002787   -20.344   < 2e-16 ***
## children        0.406545    0.016401    24.787   < 2e-16 ***
## education       0.176705    0.021531     8.207   2.27e-16 ***
## religion        0.334847    0.059133     5.663   1.49e-08 ***
## Signif. codes: 0 '***' 0.001 '**' 0.01 '*' 0.05 '.' 0.1 ' ' 1
##
```

The four covariates are significant (p <.0001) in the model. They are fixed effects. They tell about the relations in the data. Age has a negative sign, so older women were less likely to use contraceptives. However, those with more children, higher level of education, and having religious beliefs are more likely to use contraceptives. We say this since the signs on the coefficients are positive.

11.5 Three-Level Nested Logistic Regression Model with Random-Intercepts

A three-level logistic regression model with random-intercept to the model is fit to the Bangladesh data with the SAS and R programs. The response is contraceptive use. The fixed effects in the model are: age of the woman, number of living children, level of education, and whether or not woman has religious beliefs. The geographical areas (clusters – γ_{jk}) are the level-2 random effects and the divisions (δ_k)are level-3 random effects. The equation to the model is written as:

$$\text{logit}\{\mu_{ij} \mid X_{ij1}...X_{ijP}, \gamma_{jk}\delta_k\} = \beta_0 + \beta_1 X_{ij1} + ... + \beta_P X_{ijP} + \gamma_{jk} + \delta_k$$

11.5.1 Statistical Analysis of Data Using SAS Program

```
PROC GLIMMIX data=Bangladesh ;
CLASS cluster divisions;
model CUC(event='1')= age children education religion/
DIST=BINARY LINK=LOGIT DDFM=BW SOLUTION OR;
random intercept/subject=divisions solution TYPE=VC;
random intercept/subject=cluster(divisions) TYPE=VC;
covtest/WALD;
run;
```

Selected The GLIMMIX Procedure

Model Information

Class Level Information

Class	Levels	Values
districts	600	1 2 3 4 5 6 7 8 9 10 11 12 13 14 15 16 17 18 19 20 21 22 23 24 25 26 27 28 .... 594 595 596 597 598 599 600
divisions	7	Barisal Chittag Dhaka Khulna Rajshah Rangpur Sylhet

There are 600 districts (districts are nested within divisons). There are seven divisions.

Covariance Parameter Estimates

Cov Parm	Subject	Estimate	Standard Error	Z Value	Pr > Z
Intercept	Divisions	0.2066	0.1216	1.70	0.0446
Intercept	cluster(divisions)	0.1316	0.01796	7.33	<.0001

An estimate of variance random effects (divisions) is 0.2066. It is marginally significant (p <.0446). An estimate of variance random effects (clusters within divisions) is 0.1316. It is also significant (p <.0001). Therefore, accounting for the two sets of random effects in the model is a good decision.

Solutions for Fixed Effects					
Effect	Estimate	Standard Error	DF	t Value	Pr > |t|
Intercept	0.5587	0.1968	6	2.84	0.0296
age	−0.05603	0.002456	16175	−22.81	<.0001
children	0.4067	0.01527	16175	26.63	<.0001
education	0.1779	0.02059	16175	8.64	<.0001
religion	0.3746	0.05553	16175	6.75	<.0001

The four covariates are significant (p <.0001) in the model. They are fixed effects. They tell about the relations in the data. Age has a negative sign, so older women were less likely to use contraceptives. However, those with more children, higher level of education, and having religious beliefs are more likely to use contraceptive. We say this since the signs on the coefficients are positive.

Solution for Random Effects						
Effect	Subject	Estimate	Std Err Pred	DF	t Value	Pr > |t|
Intercept	divisions Barisal	0.1584	0.1818	16181	0.87	0.3838
Intercept	divisions Chittag	−0.4916	0.1796	16181	−2.74	0.0062
Intercept	divisions Dhaka	0.02810	0.1789	16181	0.16	0.8752
Intercept	divisions Khulna	0.2845	0.1803	16181	1.58	0.1145
Intercept	divisions Rajshah	0.4215	0.1804	16181	2.34	0.0195
Intercept	divisions Rangpur	0.3526	0.1807	16181	1.95	0.0510
Intercept	divisions Sylhet	−0.7535	0.1819	16181	−4.14	<.0001

A measure of the random effects for each division is provided. We note that division Sylhet does provide a significant effect as compared to others.

11.5.2 Statistical Analysis of Data Using R Program

A three-level logistic regression model with random-intercept is fit to the Bangladesh data with the R program. The response is contraceptive use. The fixed effects in the model are: age of the woman, number of living children, level of education, and whether or not the woman has religious beliefs. The areas (clusters) are the random effects, and the divisions are random effects.

```
## Load the library
library(lme4)
## Call "glmer" to the fit the three-level model
glmer.out3 = glmer(CUC~age+children+education+religion
        +(1|divisions/cluster),data=Bangladesh,family=binomial)
summary(glmer.out3)

## Generalized linear mixed model fit by maximum likelihood (Laplace
##    Approximation) [glmerMod]
##    Family: binomial ( logit )
## Formula:
## CUC ~ age + children + education + religion + (1 | divisions/
            cluster)
##       Data: Bangladesh
##
## Random effects:
##   Groups              Name         Variance   Std.Dev.
##   cluster:divisions (Intercept) 0.1367      0.3697
##   divisions           (Intercept) 0.1828      0.4276
## Number of obs: 16186, groups:    cluster:divisions, 600; divisions, 7
##
##
```

There are 600 clusters. There are seven divisions. An estimate of the variance of random effects (divisions) is 0.1828 with standard deviation 0.4276. An estimate variance of random effects (clusters within divisions) is 0.1367 with standard deviations 0.3697. These results are a little different from those obtained with SAS and STATA due to different default settings.

```
## Fixed effects:
##              Estimate Std. Error    z value Pr(>|z|)
## (Intercept)  0.570587     0.188687    3.024   0.00249     **
## age         -0.057041     0.002512  -22.708  < 2e-16      ***
## children     0.413957     0.015699   26.369  < 2e-16      ***
## education    0.180806     0.020776    8.703  < 2e-16      ***
## religion     0.380421     0.056030    6.790  1.12e-11     ***
## Signif. codes: 0 '***' 0.001 '**' 0.01 '*' 0.05 '.' 0.1 ' ' 1
##
```

The four covariates are significant (p <.0001) in the model. They are fixed effects. They tell about the relations in the data. Age has a negative sign, so older women were less likely to use contraceptives. However, those with more children, higher level of education and having religious beliefs are more likely to use contraceptives. We say this since the signs on the coefficients are positive.

11.6 Research/Questions and Comments

11.6.1 Research Questions and Answers

The following questions are revisited. The hierarchical models provide the answers:

Q1. **How much variance in contraceptive use is attributable to areas and divisions?**

Analyze the three-level hierarchical data without fixed effects; thus, one has an unconditional model. This model provides an overall estimate of the contraceptive use for women living in a typical geographical area in a typical division. It provides the amount of variability in contraceptive use due to geographic areas within divisions as well as the variability between divisions. Thus, compute the intra-class correlation coefficient (that indicates how much of the total variation in the probability of the contraceptive use test is accounted for by geographic areas and divisions). The total variation is measured as

$$var(division) + var(cluster\ within\ division) + var(level\ 1)$$

$$= [0.112 + 0.08645 + 3.29].$$

The 3.29 is the lowest level variance when fitting logistic regression (Ene, Leighton, Blue, and Bell 2015). From the ICC, it is determined that 2.5% $= \dfrac{0.112}{0.112 + 0.08645 + 3.29}$ of the variability in the contraceptive use is accounted for by the geographic areas within divisions, and 3.2% $= \dfrac{0.08645}{0.112 + 0.08645 + 3.29}$ of the variability is accounted for by the divisions, leaving 95% $= 1 -$ $(0.025 + 0.032)$ of the variability to be accounted for by the women's characteristics and other factors. Based on the p-values for each of the variance estimates, it is concluded that the variability between geographic areas

TABLE 11.2

Covariance Parameter Estimates

Cov Parm	Subject	Estimate	Standard Error	Z Value	Pr > Z
Intercept	Divisions	0.1121	0.06655	1.68	0.0460
Intercept	Cluster (divisions)	0.08645	0.01438	6.01	<.0001

Solutions for Fixed Effects

Effect	Estimate	Standard Error	DF	t Value	Pr > |t|
Intercept	0.4015	0.1282	6	3.13	0.0203

(p <.0001) and between divisions (p = 0.0460) are statistically significant, as shown in Table 11.2.

Q2. **Does the influence of age vary among geographical areas?**

This is answered with the random slope with age in that model.

	Covariance Parameter Estimates				
Cov Parm	Subject	Estimate	Standard Error	Z Value	Pr > Z
Intercept	Cluster	0.2324	0.04180	5.56	<.0001
age	Cluster	0.000066	0.000037	1.80	0.0360

The two-level logistic regression model, with random-intercept and random-slope in age, addresses the question. The age is significant across areas, so its influence varies (variance in random age is 0.000066 with p–value = 0.0360).

Q3. **What is the impact of geographical area, while controlling for women's characteristics?**

This was answered with a two-level logistic regression model with area as random-intercept.

$$\text{logit}\{\mu_{ij} \mid X_{ij1} \ldots X_{ijP}, \gamma_j\} = \beta_0 + \beta_1 X_{ij1} + \ldots + \beta_P X_{ijP} + \gamma_j$$

	Covariance Parameter Estimates				
Cov Parm	Subject	Estimate	Standard Error	Z Value	Pr > Z
Intercept	cluster	0.2931	0.02790	10.51	<.0001
	Solutions for Fixed Effects				
Effect	Estimate	Standard Error	DF	t Value	Pr > |t|
Intercept	0.6296	0.09974	599	6.31	<.0001

age	−0.05428	0.002457	15582	−22.09	<.0001
children	0.3894	0.01512	15582	25.76	<.0001
education	0.1670	0.02093	15582	7.98	<.0001
religion	0.3217	0.05771	15582	5.57	<.0001

The geographical area is significant while controlling for women's character-istics. Thus, one can conclude that there is a relationship between geographic area experience and the probability of contraceptive use. For a more intuitive interpretation, one can interpret the odds of contraceptive usage, which is 1.026; for a one-unit increase in geographic area experience, one expects to see about a 3% increase in the odds of contraceptive use.

Q4. Does the division have an impact on contraceptive use while controlling for women's characteristics?

This was answered with the three-level logistic regression model with divi-sions as random-intercepts. The covariance parameter estimates suggest that there is enough variability due to geographical areas and the divisions.

Covariance Parameter Estimates					
Cov Parm	Subject	Estimate	Standard Error	Z Value	Pr > Z
Intercept	Divisions	0.2066	0.1216	1.70	0.0446
Intercept	Cluster (divisions)	0.1316	0.01796	7.33	<.0001

The random effect for divisions was significant, though not as significant as the areas. Nevertheless, contraceptive use differed across divisions.

Q5. If one ignores the overdispersion that is present due to correlated data, what are the consequences?

Ignoring the overdispersion results in a lower standard error that is actually present. Thus, the test statistics are larger than they should be. This results in smaller p-values. One is likely to declare significance when there is not.

11.6.2 Comments

Logistic regression models with random effects are subject-specific models and, therefore, are different from marginal models. The two sets of models do not answer the same question. They resemble each other, however, as one models the mean of the binary outcome variable in the marginal model, while the other models the conditional mean in the subject-specific model.

The Hosmer-Lemeshow test is a goodness-of-fit for the standard logistic regression model. If the test shows that it is not a good fit, then one may consider fitting a logistic regression model with random-intercepts. There is no such test to fit the random coefficient models. One can test the random effects variance to be zero.

11.7 Exercises

1. Fit a logistic regression model to the contraceptive use data using the following covariates: age at first marriage, television in the household, radio in the household, religion, and family wealth.

2. Fit a logistic regression model with clusters and division as random to the contraceptive use data intercept while controlling for the following covariates: age at first marriage, television in the household, radio in the household, religion, and family wealth.

3. Of the random effects, which cluster in which divisions seems to have a positive effect on contraceptive use?

References

Blackwelder, W.C.: Equivalence trials. In: Armitage, P., Colton. T. (eds). *Encyclopedia of Biostatistics*, vol. 2, pp. 1367–1372. Wiley, New York (1998)

Dean, C.B., Nielsen, J.D.: Generalized linear mixed models: a review and some extensions. *Lifetime Data Analysis*, 13(4), 497–512 (2007)

Dobson, A.J.: *An Introduction to Generalized Linear Models*. Chapman & Hall/CRC, Boca Raton (2002)

Ene, M., Leighton, E.A., Blue, G., Bethany A.: Multilevel models for categorical data using SAS®. PROC GLIMMIX: The Basics. SAS Global Forum Paper 3430-2015. http://support.sas.com/resources/papers/proceedings15/3430-2015.pdf (2015)

Goldstein, H.: *Multilevel Statistical Models*, 3rd ed. Edward Arnold, London (2003)

Have, T.R., Kunselman, A.R., Tran, L.A.: comparison of mixed effects logistic regression models for binary response data with two nested levels of clustering. *Statistics in Medicine*, 18, 947–960 (1999)

Hu, F.B., Goldberg, J., Hedeker, D., Flay, B. R., & Pentz, M. A.: Comparison of population-averaged and subject-specific approaches for analyzing repeated binary outcomes. *American Journal of Epidemiology*, 147(7), 694–703 (1998)

Khan H.R., Shaw, E.: Multilevel logistic regression analysis applied to binary contraceptive prevalence data. *Journal of Data Science*, 9, 93–110 (2011)

Larsen, K., Petersen, J.H., Budtz-Jørgensen, E., Endahl, L.: Interpreting parameters in the logistic regression model with random effects. *Biometrics*, 56, 909–914 (2000)

Schabenberger, O.: *Introducing the GLIMMIX Procedure for Generalized Linear Mixed Models*. SAS Institute. https://support.sas.com/resources/papers/proceedings/proceedings/sugi30/196-30.pdf (2005)

Index

A

Algorithm converged, 190
American Statistical Association, 73, 95
Analytical approach, 5
ANOVA, *see* One-way analysis of
 variance model
Arithmetic mean, 22
Asymptotic test, 57
Auto regressive 1 model, 196–197

B

Bangladesh data
 hierarchical logistic regression
  models, 228–229
 standard logistic regression model,
  147–148
Bangladesh Demographic and Health
 Survey (BDHS), 147, 228
Belief measurement, 35–36
Bernoulli trial, 54
Beta (β), 13
Binary indicators, 21
Binary response, statistical model
 with one categorical factor
  homogeneity-hypotheses test,
   93–94
  independence-hypotheses test, 94
  Pearson chi-square, 94–96
  R program, data analysis, 97–98
  SAS program, data analysis, 96–97
Binomial distribution, 54
Body mass index (BMI), 19, 21, 26, 111
 vs. age, 110
 graphical distribution by gender, 19, 20
 R program, 27
 SAS program, 26–27
Bonferonni method, 87

C

Case-control studies, 12
Categorical responses, statistical model

with binary covariate, 71–73
 R program, data analysis, 75–77
 SAS program, data analysis, 73–75
parametric tests, 53–54
proportion tests, 54–55
R program, data analysis, 56
SAS program, data analysis, 55–56
Cell probability, 95
Census, 2–3
Central limit theorem (CLT), 60
Central tendency, 21–24
Chi-square distribution, 73
Class level information, 159, 190
CLT, *see* Central limit theorem
Cluster random sampling, 8
Coefficient of determination, 113–114
Cohort studies, 11–12
Cohort surveys, 11
Complex sampling design, 9
Confidence interval, 25, 48
Confounding variable, 3
Contingency table, 30–31
Continuous response, statistical model
 with binary factor, 61–62
  assumptions, 63
  population mean difference,
   parametric tests, 63
  two-sample independent
   t-statistic, 64
  two-sample t-test, 6
 with categorical and continuous
  covariates, 133–144
 with continuous and qualitative
  predictors, 124–125
 with more than two levels, 125
 R program, data analysis, 127–128
 SAS program, data analysis,
  125–127
 with continuous predictors, 128–129
 R program, data analysis, 130–131
 SAS program, data analysis,
  129–130
 with multiple factors, 120–122
 R program, data analysis, 123–124

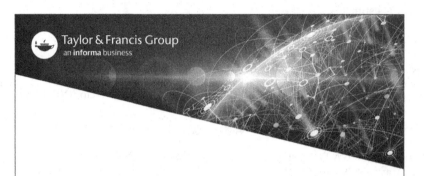

Printed in the United States
by Baker & Taylor Publisher Services